MATHEMATICAL CURIOSITIES

ALSO BY ALFRED S. POSAMENTIER
AND INGMAR LEHMANN

Magnificent Mistakes in Mathematics
The Fabulous Fibonacci Numbers
Pi: A Biography of the World's Most Mysterious Number
The Secrets of Triangles
Mathematical Amazements and Surprises
The Glorious Golden Ratio

ALSO BY ALFRED S. POSAMENTIER

The Pythagorean Theorem
Math Charmers

MATHEMATICAL CURIOSITIES

*A Treasure Trove of
Unexpected Entertainments*

ALFRED S. POSAMENTIER
INGMAR LEHMANN

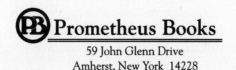

59 John Glenn Drive
Amherst, New York 14228

Published 2014 by Prometheus Books

Cover image © Bigstock
Cover design by Grace M. Conti-Zilsberger
Unless otherwise indicated, all figures and images in the text are either in the public domain or are by Alfred S. Posamentier and/or Ingmar Lehmann.

Inquiries should be addressed to
Prometheus Books
59 John Glenn Drive
Amherst, New York 14228
VOICE: 716–691–0133
FAX: 716–691–0137
WWW.PROMETHEUSBOOKS.COM
18 17 16 15 14   5 4 3 2 1

Library of Congress Cataloging-in-Publication Data

Posamentier, Alfred S.
    Mathematical curiosities : a treasure trove of unexpected entertainments / by Alfred S. Posamentier and Ingmar Lehmann.
        pages cm
    Includes bibliographical references and index.
    ISBN 978-1-61614-931-4 (pbk.) — ISBN 978-1-61614-932-1 (ebook)
1. Mathematics—Miscellanea. 2. Mathematics—Study and teaching. I. Lehmann, Ingmar. II. Title.
    QA99.P6655 2014
    510—dc23

                                                                                    2014006790

Printed in the United States of America

We dedicate this book of mathematical entertainments to our future generations so that they will be among the multitude that we hope will learn to love mathematics for its power and beauty!

To our children and grandchildren, whose future is unbounded:

Lisa, Daniel, David, Lauren, Max, Samuel, and Jack.
—Alfred S. Posamentier

Maren, Tristan, Claudia, Simon, and Miriam.
—Ingmar Lehmann

CONTENTS

ACKNOWLEDGMENTS

T he authors wish to extend sincere thanks for proofreading and extremely useful suggestions offered by Dr. Elaine Paris, emeritus professor of mathematics at Mercy College, New York. Her insight and sensitivity for the general readership has been extremely helpful. We also thank Catherine Roberts-Abel for very capably managing the production of this book, and Jade Zora Scibilia for the truly outstanding editing throughout the various phases of production. Steven L. Mitchell deserves praise for enabling us to continue to approach the general readership with yet another book demonstrating mathematical gems.

INTRODUCTION

It is unfortunate that too many people would be hard-pressed to consider anything mathematical as entertainment. Yet with this book we hope to convert the uninitiated general readership to an appreciation for mathematics—and from a very unusual point of view: through a wide variety of mathematical curiosities. These include, but certainly are not limited to, peculiarities involving numbers and number relationships, surprising logical thinking, unusual geometric characteristics, seemingly difficult (yet easily understood) problems that can be solved surprisingly simply, curious relationships between algebra and geometry, and an uncommon view of common fractions.

In order to allow the reader to genuinely appreciate the power and beauty of mathematics, we navigate these unexpected curiosities in a brief and simple fashion. As we navigate through these truly amazing representations of mathematics, we encounter in the first chapter patterns and relationships among numbers that a reader on first seeing these will think they are contrived, but they are not. It is simply that we have dug out these morsels of fantastic relationships that have bypassed most of us during our school days. It is unfortunate that teachers don't take the time to search for some of these beauties, since students in their development stages would see mathematics from a far more favorable point of view.

During its centuries of isolation, the Japanese population was fascinated with *Sangaku* puzzles, which we will admire in chapter 2 for the geometry they exhibit. It will allow us to see a curious side of geometry that may have passed us by as we studied geometry in school. We use these puzzles as a gateway to look at some other geometric manifestations.

Problem solving, as most people may recall from their school days, was presented in the form of either drill questions or carefully categorized topical problems. In the case of drill, rote memorization was expected; while in the case of topical problems, a mechanical response was too often encouraged by teachers. What was missing were the many mathematical challenges—problems in a genuine sense—that are off the beaten path, that do not necessarily fit a certain category, that can be very easily stated, and that provide the opportunity for some quite surprisingly simple solutions. These problems, provided in the third chapter, are intended to fascinate and captivate the uninitiated!

Measures of central tendency have largely been relegated to the study of statistics—as well they might be. However, when seen from a strictly mathematical point of view—algebraic and geometric—they provide a wonderful opportunity for geometrically justifying algebraic results or algebraically justifying geometric results. We do this in chapter 4, largely in the context of comparing the relative sizes of the four most popular means, or measures of central tendency, namely, the arithmetic mean (the common average), the geometric mean, the harmonic mean, and the root-mean-square.

When fractions are taught in school, they are presented largely in the context of doing the four basic arithmetic operations with them. In our last chapter, we present fractions from a completely different standpoint. First recognizing that the ancient Egyptians used only unit fractions (i.e., those in which the numerator is 1) and the fraction $\frac{2}{3}$. We will present unit fractions in a most unusual way as part of a harmonic triangle and eventually leading to Farey sequences. The reader should be fascinated to witness that fractions can be about more than just representing a quantity and being manipulated with others.

Every attempt has been made to make these curiosities as reader friendly, attractive, and motivating as possible so as to convince the general readership that mathematics is all around us and can be fun. One by-product of this book is to make the reader more quantitatively and logically aware of the world around him or her.

There are numerous examples in this treasure trove that we hope to have presented in a highly intelligible fashion so as to rekindle readers' true love for mathematics, and so that those who might have been a bit skeptical about whether curiosities that can exhibit the power and beauty of mathematics will now go forward and serve as ambassadors for the great field of mathematics. One of our goals in this book is to convince the general populace that they should enjoy mathematics and not boast of having been weak in the subject during their school years.

Chapter 1

ARITHMETIC CURIOSITIES

The curiosities found in arithmetic and in numbers, in general, are probably boundless. They range from peculiarities of certain numbers to number relationships stemming from ordinary arithmetic processes. What makes these so entertaining are the unexpected results that are sometimes inexplicable. In this chapter we will be presenting to you some of these many arithmetic and numeric oddities in mathematics. Some are clearly errors that lead to correct results, while others are correct workings of mathematics that lead to wildly unexpected results. In either case, we hope that through the mathematics alone you will be entertained without having to apply it to other fields in either the sciences or the real world. Our intent here is to demonstrate a special beauty that can make mathematics fascinating and enjoyable.

HOWLERS

In the early years of schooling we learned to reduce fractions to make them more manageable. For this there were specific ways to do it correctly. Some wise guy seems to have come up with a shorter way to reduce some fractions. Is he right?

He was asked to reduce the fraction $\frac{26}{65}$, and did it in the following way:

$$\frac{2\cancel{6}}{\cancel{6}5} = \frac{2}{5}$$

That is, he just canceled out the 6's to get the right answer. Is this procedure correct? Can it be extended to other fractions? If this is so, then we were surely treated unfairly by our elementary-school teachers, who made us do much more work. Let's look at what was done here and see if it can be generalized.

In his book *Fallacies in Mathematics*, E. A. Maxwell refers to these cancellations as "howlers":

$$\frac{1\cancel{6}}{\cancel{6}4} = \frac{1}{4} \qquad \frac{1\cancel{9}}{\cancel{9}5} = \frac{1}{5}$$

Perhaps when someone did the fraction reductions this way, and still got the right answer, it could just make you howl. This simple procedure continues to give us the correct answers: As we look at this awkward—yet easy—procedure, we could begin by reducing the following fractions to lowest terms:

$$\frac{16}{64}, \ \frac{19}{95}, \ \frac{26}{65}, \ \frac{49}{98}.$$

After you have reduced to lowest terms each of the fractions in the usual manner, one may ask why it couldn't have been done in the following way.

$$\frac{1\cancel{6}}{\cancel{6}4} = \frac{1}{4}$$

$$\frac{1\cancel{9}}{\cancel{9}5} = \frac{1}{5}$$

$$\frac{2\cancel{6}}{\cancel{6}5} = \frac{2}{5}$$

$$\frac{4\cancel{9}}{\cancel{9}8} = \frac{4}{8} = \frac{1}{2}$$

At this point you may be somewhat amazed. Your first reaction is probably to ask if this can be done to any fraction composed of two-digit numbers of this sort. Can you find another fraction (comprised of two-digit numbers) where this type of cancellation will work? You might cite $\frac{55}{55} = \frac{5}{5} = 1$ as an illustration of this type of cancellation. This will, clearly, hold true for all two-digit multiples of eleven.

For those readers with a good working knowledge of elementary algebra, we can "explain" this awkward occurrence. That is, why are the four fractions above the only ones (comprised of different two-digit numbers) where this type of cancellation will hold true?

Consider the fraction $\frac{10x+a}{10a+y}$, in which the second digit of the numerator and the first digit of the denominator match.

The above four cancellations were such that when canceling the a's the fraction was equal to $\frac{x}{y}$.

Let us explore this relationship: $\frac{10x+a}{10a+y} = \frac{x}{y}$.

$$\text{This yields } y(10x+a) = x(10a+y),$$

$$10xy + ay = 10ax + xy,$$

$$9xy + ay = 10ax \rightarrow y(9x + a) = 10ax,$$

$$\text{and so } \frac{10ax}{9x+a}.$$

At this point we shall inspect this equation. It is necessary that x, y, and a are integers since they were digits in the numerator and denominator of a fraction. It is now our task to find the values of a and x for which y will also be integral. To avoid a lot of algebraic manipulation, you will want to set up a chart that will generate values of y from $y = \frac{10ax}{9x+a}$. Remember that x, y, and a must be single-digit integers. Below is a portion of the table you will be constructing. Notice that the cases where $x = a$ are excluded, since $\frac{x}{a} = 1$.

x / a	1	2	3	4	5	6	...	9
1		$\frac{20}{11}$	$\frac{30}{12}$	$\frac{40}{13}$	$\frac{50}{14}$	$\frac{60}{15}=4$		$\frac{90}{18}=5$
2	$\frac{20}{19}$		$\frac{60}{21}$	$\frac{80}{22}$	$\frac{100}{23}$	$\frac{120}{24}=5$		
3	$\frac{30}{28}$	$\frac{60}{29}$		$\frac{120}{31}$	$\frac{150}{32}$	$\frac{180}{33}$		
4								$\frac{360}{45}=8$
⋮								
9								

Figure 1.1

This small portion of the chart (figure 1.1) already generated two of the four integral values of y; that is, when $x = 1$, $a = 6$, then $y = 4$, and when $x = 2$, $a = 6$, and $y = 5$. These values yield the fractions $\frac{16}{64}$ and $\frac{26}{65}$, respectively. The remaining two integral values of y will be obtained when $x = 1$ and $a = 9$, yielding $y = 5$, and when $x = 4$ and $a = 9$, yielding $y = 8$. These yield the fractions $\frac{19}{95}$ and $\frac{49}{98}$, respectively. This should convince you that there are only four such fractions composed of two-digit numbers, excluding two-digit multiples of 11.

Let's extend this idea and investigate whether there are fractions composed of numerators and denominators of more than two digits where this strange type of cancellation holds true. Try this type of cancellation with $\frac{499}{998}$. You should find that $\frac{499}{998} = \frac{4}{8} = \frac{1}{2}$.

A pattern is now emerging, and you may realize that

$$\frac{49}{98} = \frac{499}{998} = \frac{4999}{9998} = \frac{49999}{99998} = \frac{4}{8} = \frac{1}{2},$$

$$\frac{16}{64} = \frac{166}{664} = \frac{1666}{6664} = \frac{16666}{66664} = \frac{1}{4},$$

$$\frac{19}{95}=\frac{199}{995}=\frac{1999}{9995}=\frac{19999}{99995}=\frac{1}{5}, \text{ and}$$

$$\frac{26}{65}=\frac{266}{665}=\frac{2666}{6665}=\frac{26666}{66665}=\frac{2}{5}.$$

Enthusiastic readers may wish to justify these extensions of the original howlers. Readers who, at this point, have a further desire to seek out additional fractions that permit this strange cancellation should consider the following fractions. They should verify the legitimacy of this strange cancellation, and then set out to discover more such fractions.

$$\frac{3\cancel{3}2}{8\cancel{3}0}=\frac{32}{80}=\frac{2}{5}$$

$$\frac{3\cancel{8}5}{8\cancel{8}0}=\frac{35}{80}=\frac{7}{16}$$

$$\frac{1\cancel{3}8}{\cancel{3}45}=\frac{18}{45}=\frac{2}{5}$$

$$\frac{2\cancel{7}5}{7\cancel{7}0}=\frac{25}{70}=\frac{5}{14}$$

$$\frac{1\cancel{6}\cancel{3}}{\cancel{3}2\cancel{6}}=\frac{1}{2}$$

Aside from requiring an algebraic solution, which can be used to introduce a number of important premises in a motivational way, this topic can also provide some recreational activities. Here are some more of these "howlers."

$$\frac{48\cancel{4}}{\cancel{8}47}=\frac{4}{7} \quad \frac{\cancel{5}45}{65\cancel{4}}=\frac{5}{6} \quad \frac{\cancel{4}24}{7\cancel{4}2}=\frac{4}{7} \quad \frac{249}{996}=\frac{24}{96}=\frac{1}{4}$$

$$\frac{48\cancel{4}8\cancel{4}}{\cancel{8}4\cancel{8}47}=\frac{4}{7} \quad \frac{\cancel{5}4\cancel{5}45}{65\cancel{4}5\cancel{4}}=\frac{5}{6} \quad \frac{\cancel{4}2\cancel{4}24}{7\cancel{4}2\cancel{4}2}=\frac{4}{7}$$

$$\frac{3\cancel{2}4\cancel{3}}{4\cancel{3}2\cancel{4}}=\frac{3}{4} \quad \frac{\cancel{6}8\cancel{4}6}{86\cancel{4}\cancel{8}}=\frac{6}{8}=\frac{3}{4}$$

$$\frac{147\cancel{1}\cancel{4}}{\cancel{7}\cancel{1}468}=\frac{14}{68}=\frac{7}{34}=\frac{8\cancel{7}80\cancel{4}8}{98\cancel{7}80\cancel{4}}=\frac{8}{9}$$

$$\frac{1\cancel{4}28\cancel{5}71}{\cancel{4}28\cancel{5}713}=\frac{1}{3} \quad \frac{28\cancel{5}71\cancel{4}2}{\cancel{8}571\cancel{4}26}=\frac{2}{6}=\frac{1}{3} \quad \frac{3\cancel{4}61\cancel{5}38}{\cancel{4}61\cancel{5}384}=\frac{3}{4}$$

$$\frac{767123287}{876712328} = \frac{7}{8} \qquad \frac{3243243243}{4324324324} = \frac{3}{4}$$

$$\frac{1025641}{4102564} = \frac{1}{4} \qquad \frac{3243243}{4324324} = \frac{3}{4} \qquad \frac{4571428}{5714285} = \frac{4}{5}$$

$$\frac{4848484}{8484847} = \frac{4}{7} \qquad \frac{5952380}{9523808} = \frac{5}{8} \qquad \frac{4274514}{6428571} = \frac{4}{6} = \frac{2}{3}$$

$$\frac{5454545}{6545454} = \frac{5}{6} \qquad \frac{6923076}{9239768} = \frac{6}{8} = \frac{3}{4} \qquad \frac{4242424}{7424242} = \frac{4}{7}$$

$$\frac{5384615}{7538461} = \frac{5}{7} = \qquad \frac{2051282}{8205128} = \frac{2}{8} = \frac{1}{4} \qquad \frac{3116883}{8311388} = \frac{3}{8}$$

$$\frac{6486486}{8648648} = \frac{6}{8} = \frac{3}{4} \qquad \frac{484848484}{848484847} = \frac{4}{7}$$

This peculiarity shows how elementary algebra can be used to investigate an amusing number-theory situation. These are just some of the hidden treasures that mathematics continues to hold and that we will explore as we journey through this chapter.

A PAINTING TITLED *A DIFFICULT ASSIGNMENT*

At the end of the nineteenth century, the Russian artist Nikolai Petrovich Belsky[1] (1868–1945) produced a painting with the title *A Difficult Assignment*. In the painting (figure 1.2), we see a group of students mulling around a chalkboard, apparently frustrated with an assignment of calculating an arithmetic challenge.

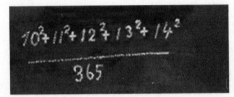

Figure 1.2a and 1.2b: Nikolai Petrovich Belsky's: *A Difficult Assignment* **(1895).**

The problem is to find the value of $\dfrac{10^2+11^2+12^2+13^2+14^2}{365}$.

Imagine trying to solve this problem without a calculator. It is certainly doable but somewhat time-consuming. However, through the amazing relationships that exist among numbers, we can see the following property that we can exploit. By partitioning the five numbers to be squared, we find that the sum of the first three squares has the same sum as the next two squares. In each case, the sum is 365, which then trivializes the original exercise.

$$\frac{(10^2+11^2+12^2)+(13^2+14^2)}{365}=\frac{365+365}{365}=2.$$

Those who recognize this pattern might also be aware of the following pattern:

$$3^2 + 4^2 = 5^2 \qquad\qquad (= 25),$$
$$10^2 + 11^2 + 12^2 = 13^2 + 14^2 \qquad\qquad (= 365),$$
$$21^2 + 22^2 + 23^2 + 24^2 = 25^2 + 26^2 + 27^2 \qquad\qquad (= 2{,}030).$$

First, you will notice that on the left side of the equals sign in each case we have one more term than we have on the right side, and the numbers being squared are consecutive.

An ambitious reader might try to find the next equation, where five squared numbers would be on the left side of the equals sign and four on the right side.

Although it might be easier to do this with a calculator, it may be more fun to look for a pattern to make our calculation even easier.

THE MAGIC OF ALGEBRA

There are times when an overwhelming arithmetic problem can be nicely simplified with some basic algebra. Let us consider one such example now. In today's world, complicated calculations are easily disposed of using a calculator. However, it is entertaining to see how using algebraic manipulation can make a very complicated calculation practically trivial.

Consider the task of finding the value of $\sqrt{1999 \cdot 2000 \cdot 2001 \cdot 2002 + 1}$. Surely, using a calculator, we can find that this cumbersome expression is equal to 4,001,999. However, it is interesting to see how we can generalize this expression to our advantage. Since the numbers being multiplied are consecutive, let's see if that gives us an advantage. We begin by letting $n = 2,000$ and express the other numbers under the radical sign in this way:

$$(n-1) \cdot n \cdot (n+1) \cdot (n+2) + 1.$$

Now for some algebraic gymnastics: by multiplying the terms of this long algebraic expression and then adding 1, we get:

$$(n-1) \cdot n \cdot (n+1) \cdot (n+2) + 1 = n^4 + 2n^3 - n^2 - 2n + 1.$$

We shall now rearrange and dismantle these terms to suit our plan to get a workable expression:

$$n^4 + 2n^3 - n^2 - 2n + 1 = n^4 + n^3 - n^2 + n^3 + n^2 - n - n^2 - n + 1.$$

This allows us to form the following product of two trinomials:

$$n^4 + n^3 - n^2 + n^3 + n^2 - n - n^2 - n + 1 = (n^2 + n - 1) \cdot (n^2 + n - 1)$$
$$= (n^2 + n - 1)^2.$$

Replacing the original term under the radical sign with its equivalent established above, we are able to simplify the expression under the radical—a perfect square—which allows us to remove the radical sign. $\sqrt{(n-1)\cdot n\cdot(n+1)\cdot(n+2)+1} = \sqrt{(n^2+n-1)^2} = \mid n^2 + n - 1\mid$.

Because we are working with natural numbers, we can conclude that

$$\sqrt{(n-1)\cdot n\cdot(n+1)\cdot(n+2)+1} = \sqrt{(n^2+n-1)^2} = n^2+n-1.$$

Therefore, when $n = 2,000$, we get

$$\sqrt{1999\cdot 2000\cdot 2001\cdot 2002+1} = 2000^2 + 2000 - 1 = 4,000,000 + 2,000 - 1 = 4,001,999,$$

which is what we expected, since it conforms to the result obtained by using a calculator.

We have seen how algebra can help us understand and also simplify arithmetic processes, and we are now ready to explore some peculiarities embedded with particular numbers. Many numerals we take for granted we view only as the quantities they represent. Here are some rather-curious numeric insights and relationships that may make you consider and appreciate these numbers differently.

THE CURIOUS NUMBER 8

The number 8, which, in the Chinese culture, is the "lucky" number, has a unique arithmetic feature: it is the only cube number that is smaller than a square number by 1. That is, $8 = 2^3 = 9 - 1 = 3^2 - 1$.

THE CURIOUS NUMBER 9

The number 9 is the only square number that is equal to the sum of the cubes of two consecutive natural numbers. That is to say, $9 = 1^3 + 2^3$.

While we are considering the sum of the cubes, we can recall the finding by the famous Swiss mathematician Leonhard Euler (1707–1783), who stated that the smallest natural number that can be expressed as the sum of the cubes of natural numbers in two ways is 1,729.

That is, $1^3 + 12^3 = 1 + 1,728 = 1,729$, and $9^3 + 10^3 = 729 + 1,000 = 1,729$.

Now returning to the number 9, we find that it can be expressed in fractional form using all ten digits exactly once, as we see in the following fractions, which are the only fractions that will give us this amazing result: $9 = \frac{95742}{10638}$, $9 = \frac{95823}{10647}$, and $9 = \frac{97524}{10836}$. However, if we allow the 0 to take a first position in any of the numbers, we get an additional three fractional equivalents to the number 9 also using all the digits exactly once.

$$9 = \frac{57429}{06381}, \quad 9 = \frac{58239}{06471}, \text{ and } 9 = \frac{75249}{08361}.$$

THE CURIOUS NUMBER 11

The number 11 is truly a curious number. According to the British king George V, the armistice in 1918 occurred at the eleventh hour of the eleventh day of the eleventh month of the year.

In the American measuring system, the number 11 appears as a factor in linear measurements as follows: $11 \cdot 20$ yards = 1 furlong, and $11 \cdot 160$ yards = 1 mile.

We should also note that the number 11 is the only palindromic prime number that has an even number of digits. Some more curiosities with the number 11 are offered here to amuse the reader further.

First, we have 11^2 equal to the sum of five consecutive powers of 3 as follows:

$$11^2 = 121 = 3^0 + 3^1 + 3^2 + 3^3 + 3^4.$$

Then we have 11^3 as the sum of the squares of three consecutive odd numbers:

$$11^3 = 1,331 = 19^2 + 21^2 + 23^2.$$

We can also express the number 11 as the sum of a square and a prime in two different ways—and it is the smallest number that has this property!

$$11 = 2^2 + 7$$
$$11 = 3^2 + 2$$

Now here is one property of 11 that is really spectacular: if we reverse the digits of any number, which is divisible by 11, the resulting number will also be divisible by 11. To demonstrate this, let us take for an example the number 135,916, which is 11 times 12,356, and reverse the digits to get 619,531, which just happens to be 11 times 56,321, clearly a multiple of 11. You may wish to try this with other multiples of 11 and entertain your friends with this peculiarity.

Here is a nice little trick involving the number 11. Take any number where no two adjacent digits have a sum greater than 9. Then multiply that number by 11, and reverse the digits of this product. Then divide this result by 11. The resulting number will be the reverse of the original number. Let us consider as an example the number 235,412, which is a number where no two adjacent digits have the sum greater than 9. When we multiply it by 11 we get $235{,}412 \cdot 11 = 2{,}589{,}532$. Reversing the digits of this product, we get 2,359,852; and then dividing it by 11, we get 214,532, which is a number whose digits are in the reverse order of the original number.

Aside from being the fifth prime number, the number 11 is also the fifth Lucas number. You may recall the *Lucas numbers* are a sequence of numbers beginning with 1 and 3, where each succeeding number is the sum of the two previous numbers, as in the following sequence: 1, 3, 4, 7, 11, 18, 29, 47, 76, 123, . . . The sequence was popularized by the French mathematician Edouard Lucas (1842–1891), who also brought popularity to the Fibonacci numbers from which he got the idea of this sequence.[2]

To the left of the Pascal triangle in figure 1.3, you will notice that the sum of the numbers in each row generates the powers of 2, while the oblique sums show the *Fibonacci numbers*, which are similar to the Lucas numbers in that they are also generated by the sum of consecutive numbers; however, this time beginning with 1 and 1 as the first two numbers. They are 1, 1, 2, 3, 5, 8, 13, 21, 34, 55, 89, . . .

We can also find the powers of 11 on the first few rows of the famous Pascal triangle as shown in figure 1.3. Up to the fifth row, the powers of 11 appear directly: $11^4 = 14{,}641 = 1 \cdot 10^4 + 4 \cdot 10^3 + 6 \cdot 10^2 + 4 \cdot 10^1 + 1 \cdot 10^0$.

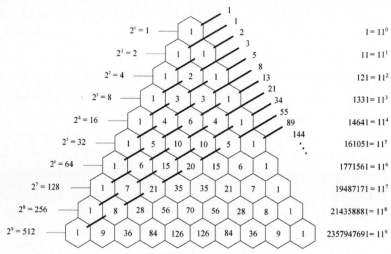

Figure 1.3

If we try to get the fifth power of 11, we notice that this row has two-digit numbers, so we will need to carry over the tens digit of each of these two-digit numbers to the next place (to the left) to get $11^5 = 161,051$.

1	5	10	10	5	1
1	5	10+1	0	5	1
1	5+1	1	0	5	1
1	**6**	**1**	**0**	**5**	**1**

You may get a better understanding of this from the following:

$$1 \cdot 10^5 + \mathbf{5} \cdot 10^4 + \mathbf{10} \cdot 10^3 + \mathbf{10} \cdot 10^2 + \mathbf{5} \cdot 10^1 + \mathbf{1} \cdot 10^0$$
$$= 1 \cdot 10^5 + 5 \cdot 10^4 + 10 \cdot 10^3 + 1 \cdot 10^3 + 0 \cdot 10^2 + 5 \cdot 10^1 + 1 \cdot 10^0$$
$$= 1 \cdot 10^5 + (5+1) \cdot 10^4 + 1 \cdot 10^3 + 0 \cdot 10^2 + 5 \cdot 10^1 + 1 \cdot 10^0$$
$$= \mathbf{1} \cdot 10^5 + \mathbf{6} \cdot 10^4 + \mathbf{1} \cdot 10^3 + \mathbf{0} \cdot 10^2 + \mathbf{5} \cdot 10^1 + \mathbf{1} \cdot 10^0$$
$$= 161,051 = 11^5.$$

Analogously, we get $11^6 = 1,771,561$ in the same way:

1	6	15	20	15	6	1
1	6	15	20 + 1	5	6	1
1	6	15 + 2	1	5	6	1
1	6 + 1	7	1	5	6	1
1	7	7	1	5	6	1

Let's discuss a very nifty way to multiply by 11. This technique always enchants the unsuspecting mathematics-phobic person, because it is so simple that it is even easier than doing it on a calculator!

The rule is very simple: *To multiply a two-digit number by 11, just add the two digits and place this sum between the two digits.*

Let's try using this technique. Suppose you wish to multiply 45 by 11. According to the rule, add 4 and 5 and place the sum between the 4 and 5 to get 495.

This does get a bit more difficult when the sum of the two digits you added results in a two-digit number. What do we do in a case like that? We no longer have a single digit to place between the two original digits. So if the sum of the two digits is greater than 9, we place the units digit between the two digits of the number being multiplied by 11 and "carry" the tens digit to be added to the hundreds digit of the product. Let's try it with 78 · 11. We begin by taking 7 + 8 = 15. We place the 5 between the 7 and 8, and add the 1 to the 7, to get [7 + 1][5][8] or 858.

You may legitimately ask if the rule also holds when 11 is multiplied by a number of more than two digits. Let's try for a larger number such as 12,345 and multiply it by 11 to see if our method still works.

Here we begin at the right-most digit and add every pair of digits, moving to the left.

$$1[1 + 2][2 + 3][3 + 4][4 + 5]5 = 135,795.$$

Recall what results when we reverse the digits of this multiple of 11 to get 597,531. When we divide this number by 11, we get 54,321. Notice how this is the reverse of the multiplier above, which was 12,345. An ambitious reader may want to determine when the multiplier will be the reverse as is the case here.

Returning now to our nifty technique for multiplying by 11, we consider a number where the sum of two consecutive digits is greater than 9. Here we use the procedure described before: place the units digit appropriately and carry the tens digit. We will do one of these here.

Multiply 456,789 by 11.

We carry the process step by step:

4[4+5][5+6][6+7][7+8][8+9]9
4[4+5][5+6][6+7][7+8][17]9
4[4+5][5+6][6+7][7+8+1][7]9
4[4+5][5+6][6+7][16][7]9
4[4+5][5+6][6+7+1][6][7]9
4[4+5][5+6][14][6][7]9
4[4+5][5+6+1][4][6][7]9
4[4+5][12][4][6][7]9
4[4+5+1][2][4][6][7]9
4[10][2][4][6][7]9
[4+1][0][2][4][6][7]9
[5][0][2][4][6][7]9
5,024,679

This rule for multiplying by 11 ought to be shared with your friends. Not only will they be impressed with your cleverness, they may also appreciate knowing this shortcut—and, above all, it will make you a good ambassador for mathematics.

At the oddest times the issue can come up to determine if a given number is divisible by 11. If you have a calculator at hand, the problem is easily solved. But that is not always the case. Besides, there is such a clever "rule" for testing for divisibility by 11 that it is worth knowing just for its charm, not to mention its utility.

The rule is quite simple: *If and only if the difference of the sums of the alternate digits is divisible by 11, then the original number is also divisible by 11.*

This may sound a bit complicated, but it really isn't. Let us take this rule one piece at a time. The sums of the alternate digits means you begin at one end of the number, taking the first, third, and fifth digits (and so on) and add them. Then, for the second sum, add the remaining (even-placed) digits. Subtract the two sums, and inspect the difference for divisibility by 11.

This may be best demonstrated through an example. We shall test 768,614 for divisibility by 11. Sums of the alternate digits are: $7 + 8 + 1 = 16$, and $6 + 6 + 4 = 16$. The difference of these two sums, $16 - 16 = 0$, which is divisible by 11. Therefore, we can conclude that 768,614 is divisible by 11.

Another example might be helpful to firm up your understanding of this procedure. To determine if 918,082 is divisible by 11, we find the sums of the alternate digits: $9 + 8 + 8 = 25$, and $1 + 0 + 2 = 3$. Their difference is $25 - 3 = 22$, which is divisible by 11, and so again we can conclude that the number 918,082 is divisible by 11.[3]

NUMBERS WHERE ALL THE DIGITS ARE ONES

Having seen some of the unusual features of the number 11, let's consider large numbers consisting of only 1s—called *repunits*.[4]

The next larger number after 11 that consists of all 1s is the number 111, and it, too, has some curious properties.

It is the third difference of two squares, and the number 1,111 is the fourth difference of two squares. We find that the progression of such differences of squares leads to repunit numbers:

$$1^2 - 0^2 = 1$$
$$6^2 - 5^2 = 11$$
$$20^2 - 17^2 = 111$$
$$56^2 - 45^2 = 1,111$$
$$156^2 - 115^2 = 11,111$$

$$556^2 - 445^2 = 111,111$$
$$344^2 - 85^2 = 111,111$$
$$356^2 - 125^2 = 111,111$$

Within this list of numbers that will result in 1s, we see a pattern emerging. Look at the second, fourth, and sixth entries. You will notice an additional pattern between the generating numbers. Each time an additional 5 and 4 is tagged onto the front of the numbers, respectively, we create another repunit. If we continue this pattern, notice what a spectacular pattern evolves.

$$6^2 - 5^2 = 11$$
$$56^2 - 45^2 = 1,111$$
$$556^2 - 445^2 = 111,111$$
$$5556^2 - 4445^2 = 11,111,111$$
$$55556^2 - 44445^2 = 1,111,111,111$$
$$555556^2 - 444445^2 = 111,111,111,111$$
$$5555556^2 - 4444445^2 = 11,111,111,111,111$$
$$55555556^2 - 44444445^2 = 1,111,111,111,111,111$$
$$555555556^2 - 444444445^2 = 111,111,111,111,111,111$$

$$\ldots$$

$$55555555555555556^2 - 44444444444444445^2$$
$$= 1,111,111,111,111,111,111,111,111,111,111,111$$

Of this list, the only prime number is 11. In fact, the next two prime numbers of all 1s are 1,111,111,111,111,111,111, and 11,111,111,111,111,111,111,111. It is quite obvious that these last two numbers will be prime in any arrangement of the digits—since they are all the same!

However, we should be aware that there are, in fact, prime numbers, where all arrangements of their digits result in another prime number. The first few of these are 11, 13, 17, 37, 79, 113, 199, and 337. You might like to find the next few such primes that create new prime numbers with each arrangement of their digits.

The story about our "curious number 11" continues as we examine the numbers generated by the difference of squares that are multiples of numbers consisting of numbers containing only 1s.

$$7^2 - 4^2 = 33 = 3 \cdot 11$$
$$67^2 - 34^2 = 3{,}333 = 3 \cdot 1{,}111$$
$$667^2 - 334^2 = 333{,}333 = 3 \cdot 111{,}111$$
$$6667^2 - 3334^2 = 33{,}333{,}333 = 3 \cdot 11{,}111{,}111$$
$$66667^2 - 33334^2 = 3{,}333{,}333{,}333 = 3 \cdot 1{,}111{,}111{,}111$$

Here is another such pattern of numbers that should be admired.

$$8^2 - 3^2 = 55 = 5 \cdot 11$$
$$78^2 - 23^2 = 5555 = 5 \cdot 1{,}111$$
$$778^2 - 223^2 = 555{,}555 = 5 \cdot 111{,}111$$
$$7778^2 - 2223^2 = 55{,}555{,}555 = 5 \cdot 11{,}111{,}111$$
$$77778^2 - 22223^2 = 5{,}555{,}555{,}555 = 5 \cdot 1{,}111{,}111{,}111$$

On further investigation of repunit numbers, we discover an interesting pattern, one in which we divide 111,111,111 by 9 to give us the number 12,345,679. Notice we have the digits in numerical order, but we are missing the digit 8. Yet when we consider the following pattern, the 8 is once again included in generating numbers consisting of only 1s.

$$
\begin{aligned}
0 \cdot 9 + 1 &= 1 \\
1 \cdot 9 + 2 &= 11 \\
12 \cdot 9 + 3 &= 111 \\
123 \cdot 9 + 4 &= 1{,}111 \\
1{,}234 \cdot 9 + 5 &= 11{,}111 \\
12{,}345 \cdot 9 + 6 &= 111{,}111 \\
123{,}456 \cdot 9 + 7 &= 1{,}111{,}111 \\
1{,}234{,}567 \cdot 9 + 8 &= 11{,}111{,}111 \\
12{,}345{,}678 \cdot 9 + 9 &= 111{,}111{,}111
\end{aligned}
$$

Don't stop now—continue this pattern to the following:
$123,456,789 \cdot 9 + 10 = 1,111,111,111$

As you can see, repunit numbers (sometimes also referred to as unit-digit numbers) seem to generate some rather-interesting relationships and patterns. Let us investigate what happens when we take the square of successive unit-digit numbers as shown in figure 1.4.

Number of 1's	n	n^2
1	1	1
2	11	121
3	111	12321
4	1111	1234321
5	11111	123454321
6	111111	1234564321
7	1111111	1234567654321
8	11111111	123456787654321
9	111111111	12345678987654321
10	1111111111	1234567900987654321

Figure 1.4

To get a better view of these repunit numbers, r_n, we will factor them into their prime factors as follows:

$r_1 = 1 = 1$ $\quad$ $r_{11} = 11111111111 = 21,649 \cdot 513,239$

$r_2 = 11 = 11$ $\quad$ $r_{12} = 111111111111 = 3 \cdot 7 \cdot 11 \cdot 13 \cdot 37 \cdot 101 \cdot 9,901$

$r_3 = 111 = 3 \cdot 37$ $\quad$ $r_{13} = 1111111111111 = 53 \cdot 79 \cdot 265,371,653$

$r_4 = 1111 = 11 \cdot 101$ $\quad$ $r_{14} = 11111111111111 = 11 \cdot 239 \cdot 4,649 \cdot 909,091$

$r_5 = 11111 = 41 \cdot 271$ $\quad$ $r_{15} = 111111111111111 = 3 \cdot 31 \cdot 37 \cdot 41 \cdot 271 \cdot 2,906,161$

$r_6 = 111111 = 3 \cdot 7 \cdot 11 \cdot 13 \cdot 37$ $\quad$ $r_{16} = 1111111111111111 = 11 \cdot 17 \cdot 73 \cdot 101 \cdot 137 \cdot 5,882,353$

$r_7 = 1111111 = 239 \cdot 4,649$ $\quad$ $r_{17} = 11111111111111111 = 2,071,723 \cdot 5,363,222,357$

$r_8 = 11111111 = 11 \cdot 73 \cdot 101 \cdot 137$ $\quad$ $r_{18} = 111111111111111111 = 3^2 \cdot 7 \cdot 11 \cdot 13 \cdot 19 \cdot 37 \cdot 52,579 \cdot 333,667$

$r_9 = 111111111 = 3^2 \cdot 37 \cdot 333,667$ $\quad$ $r_{19} = 1111111111111111111 = 1111111111111111111$

$r_{10} = 1111111111 = 11 \cdot 41 \cdot 271 \cdot 9,091$ $\quad$ $r_{20} = 11111111111111111111 = 11 \cdot 41 \cdot 101 \cdot 271 \cdot 3,541 \cdot 9,091 \cdot 27,961$

We notice that r_2 and r_{19} are prime numbers. The question then arises, are there other such repunits that are primes? The answer is yes. However, mathematicians have struggled with this question for years. For example, the German mathematician Gustav Jacob Jacobi (1804–1851) pursued the question as to whether the repunit r_{11} is a prime number. Today, a computer algebra system can answer this question in less than a second. Factoring repunit numbers is often very difficult; however, with the aid of a computer we find that the repunit r_{71} is factorable as follows and, therefore, is not a prime number.

r_{71} = 11111111111111111111111111111111111111111111111111111111111111111111111
= 241573142393627673576957439049 · 45994811347886846310221728895223034301839.

By 1930, it was known that r_2 and r_{19} (Oscar Hoppe, 1916) as well as r_{23} (Lehmer and Kraitchik, 1929) are prime numbers.[5] In 1970, a mathematics student, E. Seah, was able to demonstrate that the repunit r_{317} is also a prime number. The search for a prime number among the repunits continues. For example, in 1985 the repunit $r_{1,031}$ was discovered to be a prime by H. C. Williams and H. Dubner. Further repunits that have been identified as primes are: $r_{49,081}$ (H. Dubner, 1999), $r_{86,453}$ (L. Baxter, 2000), $r_{109,297}$ (P. Bourdelais and H. Dubner, 2007), and $r_{270,343}$ (M. Voznyy and A. Budnyy, 2007).[6] It is believed today that there are an infinite number of repunits that are prime numbers.

THE SPECIAL NUMBER TRIO: 16, 17, 18

The three numbers 16, 17, and 18 have a curious relationship. First of all, let us look at the pair of numbers 16 and 18, which have a very special relationship. Each of these numbers can represent the area of a rectangle that is numerically equal to its perimeter. That is to say that a rectangle (in this case a square) whose sides have length 4 has an area of 16 square units and a perimeter of 16 units. Analogously, a rectangle that has side lengths 3 and 6 will have an area of 18 square units and a perimeter of 18 units. These are the only two natural numbers where this property holds true.

Examining 16 individually, we find that it can be written as two base/powers interchanged as $16 = 2^4$ and $16 = 4^2$. No other number has this property![7]

Recalling triangular numbers, the number 16 can be written as the sum of triangular numbers[8] in two ways. It is the smallest such square number that has this characteristic. $16 = 6 + 10 = 1 + 15$.

The Pythagoreans were enchanted with these two numbers and despised the number 17, which separated them. However, the number 17 has some properties that give it importance as well. The number 17 is the seventh prime number, and 17 generates the sixth Mersenne number,[9] 131,071. Furthermore, 17 is the sum of the first four prime numbers: $2 + 3 + 5 + 7 = 17$. Perhaps its most famous manifestation is that it represents the number of sides of a regular polygon that can be constructed with straightedge and compass—which the famous German mathematician Carl Friedrich Gauss (1777–1855) was very proud to have proved in his early years. He was so proud of this discovery that it was ultimately constructed on his tombstone.

The number 17 also has some strange characteristics. For example, $17^3 = 4,913$ and the sum of the digits of this number is: $4 + 9 + 1 + 3 = 17$. By the way, the only other numbers that share this characteristic are: 1, 8, 18, 26, and 27. You might want check this to be convinced of this property. We offer one here: $26^3 = 17,576$, and $1 + 7 + 5 + 7 + 6 = 26$.

Some prime numbers, when their digits are reversed, also deliver prime numbers. As you can see from the list of the first few of these, the number 17 is one of these unusual numbers:

13, 17, 31, 37, 71, 73, 79, 97, 107, 113, 149, 157, . . .

Let us now inspect what can make the number 18 stand out.

We should first notice that similar to the number 17, the number 18, when spoken quickly, is often confused with the number 80. In our society, the number 18 represents the number of holes on a complete golf course, the number of wheels on a trailer truck, and the minimum voting age in many states.

Yet another curiosity can be seen with the two eighteen-letter words *conservationalists* and *conversationalists* are anagrams of each other. They are the longest pair of anagrams in the English language—if scientific words are excluded.[10]

From ancient times the number 18 has enjoyed a peculiar popularity among those who understand the Hebrew language. For centuries, Hebrew scholars have used a procedure called *gematria* to analyze the scriptures. This technique involves having the letters of a word in the Hebrew language take on their numerical equivalent. When one does this with the number 18 as expressed with Hebrew characters, it looks like this חי . In the Hebrew language, when seen as a word, these two letters spell out the word *life* and often it is seen as a good luck charm. In the Chinese culture, the number 18 represents a word that means that someone will prosper.

However, the number 18 also holds some interesting mathematical properties. For example, it is the only number that is twice the sum of its digits. Yet, when we look at this phenomenon, we can extend this unusual fact as follows: $18 = 9 + 9$, and its reversal $81 = 9 \cdot 9$, which is the square of the sum of its digits. We can continue this pattern by inserting a 9 between the two digits of 18 and get the following: $198 = 99 + 99$, and again reversing the digits: $891 = 9 \cdot 99$. While on the topic, we can see that $18 + 81 = 99$, and $9 + 9 = 18$. However, we can extend this strange property as shown here.

$$
\begin{array}{llll}
18 = & 9+9 & 81 = & 9 \cdot 9 \\
198 = & 99+99 & 891 = & 9 \cdot 99 \\
1998 = & 999+999 & 8991 = & 9 \cdot 999 \\
19998 = & 9999+9999 & 89991 = & 9 \cdot 9999
\end{array}
$$

. . . and it continues!

Another property held by the number 18 is seen when we take 18 to the third and fourth powers and inspect the two products. We will find that each of the digits from 0 to 9 was used exactly once:

$18^3 = 5,832$, and $18^4 = 104,976$.

Taking this a step further, one may have noticed that the sum of the digits of $18^3 = 5,832$, whose digit sum is $5 + 8 + 3 + 2 = 18$. This, in and of itself, would be quite spectacular; however, this can be even further extended, when we look at the sixth and seventh powers of 18. Once again, these powers of 18 yield numbers where the sum of their digits is 18. That is, $18^6 = 34{,}012{,}224$, and $3 + 4 + 0 + 1 + 2 + 2 + 2 + 4 = 18$; as well as $18^7 = 612{,}220{,}032$, and $6 + 1 + 2 + 2 + 2 + 0 + 0 + 3 + 2 = 18$.

Just to take this to a "higher" level, we find that
$18^{18} = 39{,}346{,}408{,}075{,}296{,}537{,}575{,}424$, and
$3 + 9 + 3 + 4 + 6 + 4 + 0 + 8 + 0 + 7 + 5 + 2 + 9 + 6 + 5 + 3 + 7 + 5 + 7 + 5 + 4 + 2 + 4 = 108$.

We can also have some fun with the number 18. Begin by taking any three-digit number whose digits are all different, and then arrange them in order to form the largest number and then the smallest number. Subtract the smaller number from the larger number, and you will find that this answer will be a number, the sum of whose digits is 18.

Let's consider an example. We shall select the number 584. We then write the smallest number with these digits, 458, and then the largest number with these digits, 854. Subtracting these numbers, $854 - 458$, we get 396. The sum of the digits of this number $(3 + 9 + 6)$ is 18. Try it with some other three-digit number to convince yourself of this property. This makes a great impression at any dinner party you attend!

We can also look at the eighteenth Fibonacci number[11] and show that it is equal to the sum of the cubes of four consecutive numbers, as shown here: $F_{18} = 2{,}584 = 7^3 + 8^3 + 9^3 + 10^3$.

You may recall that 18 is also the sixth Lucas number.[12] You might want to look for other such surprise appearances of this popular number 18.

THE CURIOUS NUMBER 30

The number 30 is the sum of the first four square numbers:
$1^2 + 2^2 + 3^2 + 4^2 = 1 + 4 + 9 + 16 = 30$.

The number 30 is also the largest number where all the co-primes (numbers that are relatively prime, i.e., their only common factor is 1) smaller than itself (except for 1) are prime numbers. They are: 7, 11, 13, 17, 19, 23, and 29. The other numbers that have this property are 3, 4, 6, 8, 12, 18, 24, so you can see that 30 is the largest number having this property.

In 1907, a German student, H. Bonse, developed an elementary proof[13]—without calculus—of this fact about 30. The proof can also be found in the book *The Enjoyment of Mathematics: Selections from Mathematics for the Amateur*, by Rademacher and Toeplitz.[14]

THE CURIOUS NUMBER 37

We know that the number 37 is a prime number, but it also has some rather-unique properties that we will consider here. Let us call the sum of the squares of the digits of a number n as $Q^2(n)$. So for the number 37 we have $Q^2(37) = 3^2 + 7^2 = 58$. This, so far, is not very impressive, but if we now subtract the product of the digits from this number, we will get (surprisingly) $Q^2(37) - 3 \cdot 7 = 58 - 21 = 37$.

Or put another way, $Q^2(37) = 37 + 3 \cdot 7$. This a rather-amazing property, such that one gets to wonder if there are other two-digit numbers that have this same property. In general terms, $n = \overline{ab}$, where $a, b \in \{0, 1, 2, 3 \ldots, 9\}$ and $a \neq 0$; keep in mind that $\overline{ab}$ is a two-digit number in the decimal system, such that $Q^2(\overline{ab}) - a \cdot b = \overline{ab}$. Another way of expressing this is: $Q^2(\overline{ab}) = \overline{ab} + a \cdot b$.

To answer our question about there being numbers other than 37 that have this property, we consider the general case that appears as $a^2 + b^2 - a \cdot b = 10a + b$. With the aid of a computer we find that there is only one other number that shares this property with 37, and that is 48. Since for $n = 48$, $Q^2(48) - 4 \cdot 8 = 4^2 + 8^2 - 4 \cdot 8 = 16 + 64 - 32 = 48$.

An expected question can arise, namely, are there any three-digit numbers that have this property? Or is it possible to have for $n = \overline{abc}$, where $a, b, c \in \{0, 1, 2, 3, \ldots, 9\}$ and $a \neq 0$, and $\overline{abc}$ is a number in the decimal system that $Q^2(\overline{abc}) - a \cdot b \cdot c = \overline{abc}$, or $Q^2(\overline{abc}) = \overline{abc} + a \cdot b \cdot c$? Once again, with the help of the computer, we learn that there are no three-digit numbers that have this property.

The number 37 continues to impress us with yet a further extension of these properties. This time we will consider the sum of the cubes of the digits of a number n, noted as $Q^3(n)$, where $Q^3(n) = Q^3(\overline{ab}) = a^3 + b^3$.

Let's consider the product of the number 37 and the sum of the digits of our number 37:

$$37 \cdot Q(37) = 37 \cdot (3 + 7) = 37 \cdot 10 = 370.$$

The sum of the cubes of the digits is: $Q^3(37) = 3^3 + 7^3 = 27 + 343 = 370$.

This would indicate that for our number 37, the general case would look like this: $\overline{ab} \cdot Q(\overline{ab}) = Q^3(\overline{ab})$.

Once again, we ask if this is unique for two-digit numbers, or if there are other such two-digit numbers sharing this property. With the aid of a computer we get the answer: yes, there is one other case where this relationship holds true, namely, once again, for the number 48, which was obtained by solving the equation:

$(10a + b) \cdot (a + b) = a^3 + b^3$, put another way, $10a^2 + 11ab + b^2 = a^3 + b^3$.

Therefore, for $n = 48$, we get $48 \cdot Q(48) = 48 \cdot (4 + 8) = 48 \cdot 12 = 576$, and $Q^3(48) = 4^3 + 8^3 = 64 + 512 = 576$.

Can this be extended to three-digit numbers? Will $n = \overline{abc}$, where a, b, $c \in \{0, 1, 2, 3, \ldots, 9\}$ and $a \neq 0$, that $\overline{abc} \cdot Q(\overline{abc}) = Q^3(\overline{abc})$? Sadly, the computer search tells us there is no three-digit number that satisfies this property.

However, if we modify our "requirement" to a three-digit number $n = \overline{abc}$, where $a, b, c \in \{0, 1, 2, 3, \ldots, 9\}$ and $a \neq 0$, so that $\overline{abc} \cdot (\overline{ab} + c) = \overline{ab}^3 + c^3$, we find that—again with the help a computer—there are four three-digit numbers that enjoy this property, for $n = 100, 111, 147,$ and 148:

$100 \cdot (10 + 0) = 100 \cdot 10 = 1000$ and $10^3 + 0^3 = 1000 + 0 = 1000$,
$111 \cdot (11 + 1) = 111 \cdot 12 = 1332$ and $11^3 + 1^3 = 1331 + 1 = 1332$,
$147 \cdot (14 + 7) = 147 \cdot 21 = 3087$ and $14^3 + 7^3 = 2744 + 343 = 3087$,
$148 \cdot (14 + 8) = 148 \cdot 22 = 3256$ and $14^3 + 8^3 = 2744 + 512 = 3256$.

An ambitious reader may wish to seek other such relationships.

THE CURIOSITY OF THE NUMBER 72

We introduce the famous "Rule of 72." It states that—roughly speaking—*money will double in* $\frac{72}{r}$ *years, when it is invested at an annually compounded interest rate of r percent.* So, for example, if we invest money at an 8 percent compounded annual interest rate, it will double its value in $\frac{72}{r} = n$ years. Since $\frac{72}{r} = n$, therefore, $n \cdot r = 72$. To investigate why, or if, this really works, we consider the compound-interest formula: $A = P\left(1+\frac{r}{100}\right)^n$, where A is the resulting amount of money and P is the principal invested for n interest periods at r percent annually.

We need to investigate what happens when $A = 2P$. When we apply the above equation, we get the following:

$$2 = \left(1+\frac{r}{100}\right)^n, \tag{1}$$

It then follows that $n = \dfrac{\log 2}{\log\left(1+\dfrac{r}{100}\right)}.$ (2)

Let us make a table of (rounded) values (figure 1.5) from the above equation:

r	n	n · r
1	69.66071689	69.66071689
3	23.44977225	70.34931675
5	14.20669908	71.03349541
7	10.24476835	71.71337846
9	8.043231727	72.38908554
11	6.641884618	73.0607308
13	6.641884618	73.72842319
15	4.959484455	74.39226682

Figure 1.5

If we take the arithmetic mean (the average) of the *nr* values, we get 72.04092673, which is quite close to 72, and so our "Rule of 72" seems to be a very close estimate for doubling money at an annual interest rate of r percent for n interest periods.

An ambitious reader might try to determine a "rule" for tripling and quadrupling money in a manner similar to the way we dealt with the doubling of money. The above equation (2) for k-tupling would be

$$n = \frac{\log k}{\log\left(1 + \dfrac{r}{100}\right)},$$

which for $r = 8$, gives the value for $n = 29.91884022$ (log k). Thus $nr = 239.3507218 \log k$, which for $k = 3$ (the tripling effect) gives us $nr = 114.1821673$.

We could then say that for tripling money we would have a "Rule of 114."

However far one wishes to explore this topic, the important issue here is that the popular "Rule of 72" provides a useful application for investing funds and gives us a curiosity of the number 72 to consider.

THE AMAZING NUMBER 193,939

We now present the truly remarkable prime number 193,939. Were we to write it in reverse, we would get 939,391, also a prime number. By considering various transformations of this number, we get prime numbers, as shown here:

193,939,
939,391,
393,919,
939,193,
391,939, and
919,393.

π CURIOSITIES

There are many curiosities involving the number $\pi = 3.14159265358979$.
The interested reader is referred to a book that covers many of them: *Pi:
A Biography of the World's Most Mysterious Number* (Prometheus Books,
2003).[15] However, we will provide two such curiosities here. First, we will
show one from the Bible.

We mentioned earlier that there is a time-honored procedure, gematria,
by which scholars have analyzed the ancient Hebrew scriptures for centuries. One always relishes the notion that a hidden code can reveal long-lost
secrets. Such is the case with the common interpretation of the value of π
in the Bible. Until rather recently, most books on the history of mathematics traced the value of π to its earliest appearance in the Bible. When
one reads the sentence describing the pool, or fountain, in King Solomon's
courtyard as 30 cubits around and 10 cubits as the diameter, yielding
a value of π equal to 3; (that is, $\pi = \frac{30}{10} = 3$), one concludes that this was the
value of π in biblical times. Yet there are two places in the Bible where this
sentence appears, identical in every way, except for one word, spelled differently in the two citations. This description of the pool or fountain in
King Solomon's courtyard is found in 1 Kings 7:23, and in 2 Chronicles
4:2, and it reads as follows:

> *And he made the molten sea of ten cubits from brim to brim,*
> *round in compass, and the height thereof was five cubits; and a*
> *line of thirty cubits did compass it round about.*

A late eighteenth-century rabbi, Elijah of Vilna (1720–1797), one of
the great modern biblical scholars who earned the title "Gaon of Vilna"
(meaning brilliance of Vilna), came up with a remarkable discovery, one
that could make most history of mathematics books faulty, if they still say
that the Bible approximated the value of π as 3. Elijah of Vilna noticed that
the Hebrew word of "line measure" was written differently in each of the
two biblical passages mentioned above.

In 1 Kings 7:23 it was written as קוה, whereas in 2 Chronicles 4:2 it was written as קו. Elijah applied the biblical-analysis technique gematria, in which the Hebrew letters are given their appropriate numerical values according to their sequence in the Hebrew alphabet, to the two spellings of the word for "line measure," and he found the following.

The letter values are: ק = 100, ו = 6, and ה = 5. Therefore, the spelling for "line measure" in 1 Kings 7:23 is קוה = 5 + 6 + 100 = 111, while in 2 Chronicles 4:2 the spelling קו = 6 + 100 = 106. He then took the ratio of these two values: $\frac{111}{106} = 1.0472$ (to four decimal places), which he considered the necessary correction factor. When it is multiplied by 3, which was believed to be the value of π stated in the Bible, one gets 3.1416, which is π correct to four decimal places! "Wow!!!" is the usual reaction. Such accuracy is quite astonishing for ancient times. To appreciate how accurate this four-place approximation of π is, take a string to measure the circumference and diameter of several circular objects and find their quotient. You will most likely not even get near this four-place accuracy. Moreover, to really push the point of the high degree of accuracy of four decimal places, chances are if you took the average of several such π measurements, you probably still wouldn't arrive at this four-place level of accuracy.

The American mathematics-recreationalist Martin Gardner[16] mentions that if we remove the letters of our alphabet that are symmetric about a vertical line through their midpoint (see figure 1.6)—namely, A, H, I, M, O, T, U, V, W, X, and Y—we are left with consecutive groups beginning with J, as 3-(J, K, L), 1-(N), 4-(P, Q, R, S), 1-(Z) and 6-(B, C, D, E, F, G). These group sizes give us the value of π to four-place accuracy, that is, 3.1416.

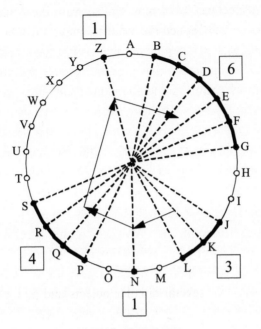

Figure 1.6

There are probably endless places in everyday life as well as in mathematics where the value of π appears. We refer the reader to the book cited above for a plethora of such applications.

Φ CURIOSITIES

One of the most ubiquitous concepts/numbers in mathematics is known as the golden ratio. Once again, this number has been celebrated by many books dedicated to it. We recommend one such book, *The Glorious Golden Ratio* (Prometheus Books, 2011).[17] The name probably stems from the fact that a rectangle whose side lengths determine the golden ratio is said to be the most beautifully shaped rectangle. This has been borne out through research by psychologists at the end of the nineteenth century. Essentially, this ratio is defined for rectangle of length l and width w as $\phi = \frac{l}{w} = \frac{w+l}{l} = \frac{\sqrt{5}+1}{2} \approx 1.6180$. This ratio appears in nature, architecture, art, and many places in science and mathematics.

Although an unsuspecting reader may find some of these appearances to be artificial, they nonetheless exist, and their interpretation is left to the beholder. Suffice it to say, investigating this ratio is a worthwhile journey into other areas of mathematics as well as our general culture. As a silly, or perhaps just simple example, you may notice that when you see a clock or watch advertisement in a newspaper, the time shown on the face of the clock is generally about 10:10. Is it a coincidence that the angle formed by the hands of the clock is approximately that of the angle formed by the diagonals of the golden rectangle? This is for the reader to decide.

THE PECULIARITY OF THE NUMBER 196

In simple terms, we can see that the number $196 = 14^2$, while the number $169 = 13^2$, just a slight rearrangement of the digits. However, the real fame that the number 196 enjoys in mathematics is that it is the smallest number that through the reversal process does not produce a palindromic number. (At least to date it has not produced a palindromic number.) Let's see what this all means. However, we must first make a slight diversion into palindromic numbers.

Palindromic numbers are those that read the same in both directions. This leads us to consider that dates can be a source for some symmetric inspection. For example, the year 2002 is a palindrome, as is 1991.[18] There were several dates in October 2001 that appeared as palindromes when written in American style: 10/1/01, or 10/22/01, and others. In February, Europeans had the ultimate palindromic moment at 8:02 p.m. on February 20, 2002, since they would have written it as 20.02, 20-02-2002. It is a bit thought provoking to come up with other palindromic dates.

As we have seen earlier, the first four powers of 11 are palindromic numbers:

$$11^1 = 11,$$
$$11^2 = 121,$$
$$11^3 = 1,331, \text{ and}$$
$$11^4 = 14,641.$$

A palindromic number can either be a prime number or a composite number. For example, 151 is a prime palindrome and 171 is a composite palindrome. Yet, with the exception of 11, a palindromic prime must have an odd number of digits.

Perhaps most interesting is to see how a palindromic number can be generated from any given number. All we need do is to add continually a number to its reversal (i.e., the number written in the reverse order of digits) until a palindrome results. For example, a palindrome can be reached with a single addition, such as with the starting number 23, we have 23 + 32 = 55, a palindrome.

Or it might take two steps to reach a palindrome, such as with the starting number 75:

$$75 + 57 = 132, \quad 132 + 231 = 363, \text{ a palindrome.}$$

Or it might take three steps, such as with the starting number 86:

$$86 + 68 = 154, 154 + 451 = 605, \quad 605 + 506 = 1,111, \text{ a palindrome.}$$

The starting number 97 will require six steps to reach a palindrome, while the number 98 will require twenty-four steps to reach the palindromic number 8,813,200,023,188.

Now we come back to the number 196. Many attempts have been made to use this rule of reverse-and-add with the number 196 to eventually reach a palindromic number. So far this is *never* been accomplished. As a matter of fact, the number 196 is the smallest number that (to the best of our current knowledge) does not generate a palindromic number through the reverse-and-add rule. The next few numbers that do not generate a palindromic number with this procedure are: 196, 295, 394, 493, 592, 689, 691, 788, 790, 879, 887, . . .

There are some lovely patterns when dealing with palindromic numbers. For example, numbers that yield palindromic cubes are palindromic themselves, as we show here:

$$11^3 \quad = \quad 1{,}331$$
$$101^3 \quad = \quad 1{,}030{,}301$$
$$1001^3 \quad = \quad 1{,}003.003{,}001$$
$$10001^3 \quad = \quad 1{,}000{,}300{,}030{,}001$$

. . . and this continues on.

THE RUTH-AARON NUMBERS

For many years the goal for most home-run-hitting baseball players was to reach, or surpass, the longtime record for career home runs set by Babe Ruth at 714 home runs. On April 8, 1974, the Atlanta Braves' slugger Hank Aaron hit his 715 career home run. This prompted the American mathematician Carl B. Pomerance (1944–) to popularize the notion (through the suggestion of one of his students) that these two numbers 714 and 715 are related by dint of the fact that they are two consecutive numbers whose prime factor sums are equal:

$$714 = 2 \cdot 3 \cdot 7 \cdot 17, \quad \text{and} \quad 2 + 3 + 7 + 17 = 29,$$
$$715 = 5 \cdot 11 \cdot 13, \quad \text{and} \quad 5 + 11 + 13 = 29.$$

We can extend the list of Ruth-Aaron numbers if we consider only numbers with *distinct* prime factors. In this case we have the following pairs: (5, 6), (24, 25), (49, 50), (77, 78), (104, 105), (153, 154), (369, 370), (492, 493), (714, 715), (1,682, 1,683), (2,107, 2,108).

If multiplicities are not counted (so that a factor of 2^3 counts only a single 2), we get the following pair:

$$(24, 25): \quad 24 = 2 \cdot 2 \cdot 2 \cdot 3, \quad \text{and } 2 + 3 = 5,$$
$$25 = 5 \cdot 5, \quad \text{and } 5 = 5.$$

If we consider repeating prime factors, then the following numbers would also qualify as Ruth-Aaron numbers: (5, 6), (8, 9), (15, 16), (77, 78), (125, 126), (714, 715), (948, 949), (1330, 1331).

$$(24, 25): \quad 8 = 2 \cdot 2 \cdot 2, \quad \text{and } 2 + 2 + 2 = 6,$$
$$9 = 3 \cdot 3, \quad \text{and } 3 + 3 = 6.$$

There are also pairs of unusual Ruth-Aaron numbers where the sums of all the factors are equal and where the sums of the factors, without repetition of each of the factors, are equal. One such pair is: (7,129,199; 7,129,200), where $7,129,199 = 7 \cdot 11^2 \cdot 19 \cdot 443$, and $7,129,200 = 2^4 \cdot 3 \cdot 5^2 \cdot 13 \cdot 457$.

Without repeating the factors of each of the numbers:
$$7 + 11 + 19 + 443 = 2 + 3 + 5 + 13 + 457 = 480.$$

The sums taking into account the repetition of the factors:
$$7 + 11 + 11 + 19 + 443 = 2 + 2 + 2 + 2 + 3 + 5 + 5 + 13 + 457 = 491.$$

We can extend the Ruth-Aaron pairs to consider Ruth-Aaron triplets, that is, three consecutive numbers where the sums of the factors are equal. Here is one such triplet: (89,460,294; 89,460,295; 89,460,296)
$$89,460,294 = 2 \cdot 3 \cdot 7 \cdot 11 \cdot 23 \cdot 8,419$$
$$89,460,295 = 5 \cdot 4,201 \cdot 4,259$$
$$89,460,296 = 2 \cdot 2 \cdot 2 \cdot 31 \cdot 43 \cdot 8,389$$

The sums of the factors are as follows:
$$2 + 3 + 7 + 11 + 23 + 8,419 = 5 + 4,201 + 4,259 = 2 + 31 + 43 + 8,389 = 8,465.$$
This is the first example without repetition!

Another such Ruth-Aaron triplet is the following: (151,165,960,539; 151,165,960,540; 151,165,960,541)
$$151,165,960,539 = 3 \cdot 11 \cdot 11 \cdot 83 \cdot 2,081 \cdot 2,411$$
$$151,165,960,540 = 2 \cdot 2 \cdot 5 \cdot 7 \cdot 293 \cdot 1,193 \cdot 3,089$$
$$151,165,960,541 = 23 \cdot 29 \cdot 157 \cdot 359 \cdot 4,021$$

The sums of the factors are:
$$3 + 11 + 83 + 2,081 + 2,411 = 2 + 5 + 7 + 293 + 1,193 + 3,089$$
$$= 23 + 29 + 157 + 359 + 4,021 = 4,589.$$

Here we have yet another Ruth-Aaron triplet, but with repetition of the factors: (417,162; 417,163; 417,164)
$$417,162 = 2 \cdot 3 \cdot 251 \cdot 277$$
$$417,163 = 17 \cdot 53 \cdot 463$$
$$417,164 = 2 \cdot 2 \cdot 11 \cdot 19 \cdot 499$$

The sums of the prime factors are:

$2 + 3 + 251 + 277 = 17 + 53 + 463 = 2 + 2 + 11 + 19 + 499 = 533.$

The last Ruth-Aaron triplet so far to have been found is: (6,913,943,284; 6,913,943,285; 6,913,943,286)

$$6,913,943,284 = 2 \cdot 2 \cdot 37 \cdot 89 \cdot 101 \cdot 5,197$$
$$6,913,943,285 = 5 \cdot 283 \cdot 1,259 \cdot 3,881$$
$$6,913,943,286 = 2 \cdot 3 \cdot 167 \cdot 2,549 \cdot 2,707$$

The sums of the prime factors are:

$$2 + 2 + 37 + 89 + 101 + 5,197 = 5 + 283 + 1,259 + 3,881$$
$$= 2 + 3 + 167 + 2,549 + 2,707 = 5,428.$$

Who would have thought that the two home-run kings would make a contribution to mathematics? By the way, we should remind ourselves that this discussion began with 714 and 715. We take the sum of these numbers $714 + 715 = 1,429$, which is a prime number—as is its reversal 9,241. Furthermore, other arrangements of this number are also prime numbers: 4,219; 4,129; 9,412; 1,249.

Interestingly, Carl B. Pomerance found that up to the number 20,000 there are twenty-six pairs of Ruth-Aaron numbers, the largest of which is (18,490; 18,491). One of the most famous mathematicians of the twentieth century, Paul Erdös (1913–1996), proved that there are an infinite number of Ruth-Aaron numbers. Feel free to search for others!

SOME CURIOSITIES ABOUT LARGE NUMBERS

Consider the number $6,666^2 = 44,435,556$. If we separate this number into two parts, each four digits long, and then add them, we get $4,443 + 5,556 = 9,999$. This will also work with the number $3,333^2 = 11,108,889$. Again, separating this number into two parts, and then adding the two parts gives us $1,110 + 8,889 = 9,999$.

Yet a more-curious result is arrived at when we use this technique on the following number: $7,777^2 = 60,481,729$, which then gives us: $6,048 + 1,729$

= 7,777, which is the same as the base we began with, unlike the previous examples that didn't result in a number related to the original number.

The numbers need not be repetitive digits to be interesting, since 297 is also a number that has this property: $297^2 = 88,209$, and $88 + 209 = 297$.

Such a number is called a *Kaprekar number*, named after the Indian mathematician Dattaraya Ramchandra Kaprekar (1905–1986), who discovered such numbers. Some other Kaprekar numbers are shown in the table of figure 1.7.

Kaprekar Number	Square of the Number			Decomposition of the Number
1	1^2	=	1	$1 = 1$
9	9^2	=	81	$8 + 1 = 9$
45	45^2	=	2,025	$20 + 25 = 45$
55	55^2	=	3,025	$30 + 25 = 55$
99	99^2	=	9,801	$98 + 01 = 99$
297	297^2	=	88,209	$88 + 209 = 297$
703	703^2	=	494,209	$494 + 209 = 703$
999	999^2	=	998,001	$998 + 001 = 999$
2,223	$2,223^2$	=	4,941,729	$494 + 1,729 = 2,223$
2,728	$2,728^2$	=	7,441,984	$744 + 1,984 = 2,728$
4,879	$4,879^2$	=	23,804,641	$238 + 04,641 = 4,879$
4,950	$4,950^2$	=	24,502,500	$2,450 + 2,500 = 4,950$
5,050	$5,050^2$	=	25,502,500	$2,550 + 2,500 = 5,050$
5,292	$5,292^2$	=	28,005,264	$28 + 005,264 = 5,292$
7,272	$7,272^2$	=	52,881,984	$5,288 + 1,984 = 7,272$
7,777	$7,777^2$	=	60,481,729	$6,048 + 1,729 = 7,777$
9,999	$9,999^2$	=	99,980,001	$9,998 + 0,001 = 9,999$
17,344	$17,344^2$	=	300,814,336	$3,008 + 14,336 = 17,344$
22,222	$22,222^2$	=	493,817,284	$4,938 + 17,284 = 22,222$

Figure 1.7

More Kaprekar numbers are:

38,962; 77,778; 82,656; 95,121; 99,999; 142,857; . . . ; 538,461; 857,143; . . . .

We also have such things as Kaprekar triples, which behave as follows:

$$45^3 = 91,125 \text{ with } 9 + 11 + 25 = 45.$$

Other Kaprekar triples are 1; 8; 10; 297; 2,322.

Previously, we showed that 297 is a Kaprekar number, and now we show how it is also a Kaprekar triple:

$$297^3 = 26,198,073, \text{ and } 26 + 198 + 073 = 297.$$

While we are dealing with discoveries made by Kaprekar, we can admire the *Kaprekar constant*, 6,174. This constant arises when one takes any four-digit number and forms the largest and the smallest number from these digits, and then subtracts these two new numbers. Continuing this process will always eventually result in the number 6,174. When the number 6,174 is arrived at, we continue the process of creating the largest and the smallest number, and then taking their difference (7,641 − 1,467 = 6,174), we notice that we get back to 6,174. This is, therefore, called the Kaprekar constant. To demonstrate this with an example, we will carry out this process with the number 2,303 as follows:

- The largest number formed with these digits is: 3,320.
- The smallest number formed with these digits is: 0,233.
- The difference is: 3,087.
- The largest number formed with these digits is: 8,730.
- The smallest number formed with these digits is: 0,378.
- The difference is: 8,352.
- The largest number formed with these digits is: 8,532.
- The smallest number formed with these digits is: 2,358.
- The difference is: 6,174.
- The largest number formed with these digits is: 7,641.
- The smallest number formed with these digits is: 1,467.
- The difference is: 6,174.

And so the loop is formed, since you continue to get the number 6,174.

Remember, all this began with an arbitrarily selected number, and it will always end up with the number 6,174, which then gets you into an endless loop (i.e., continuously getting back to 6,174).

Another curious property of 6,174 is that it is divisible by the sum of its digits: $\dfrac{6174}{6+1+7+4}=\dfrac{6174}{18}=343$.

SOME NUMBER PECULIARITIES

Consider the addition 192 + 384 = 576. You may ask, what is so special about this addition? Look at the outside digits (bold): **192** + **384** = **576**. They are in numerical sequence left to right (1, 2, 3, 4, 5, 6), and then from the right, move to the left to get the rest of the nine digits (7, 8, 9). You might have also noticed that the three numbers are:

$$192 = 1 \cdot 192,$$
$$384 = 2 \cdot 192,$$
$$576 = 3 \cdot 192.$$

Another strange result happens when we subtract the symmetric numbers, which consist of the digits in numerical order with reverse of each other: 987,654,321 − 123,456,789 to get 864,197,532. This symmetric subtraction used each of the nine digits from 1 to 9 exactly once in each of the numbers being subtracted, and surprisingly resulted in a difference that also used each of the digits exactly once.

Here are a few more such strange calculations, where on either side of the equals sign all nine digits are represented exactly once: 291,548,736 = 8 · 92 · 531 · 746, and 124,367,958 = 627 · 198,354 = 9 · 26 · 531,487.

Another example of a calculation where all nine digits (excluding 0) are used exactly once (not counting the exponents) is $567^2 = 321,489$. The same goes for the following arithmetic calculation: $854^2 = 729,316$. These are apparently the only two squares that result in a number that allows all the digits to be represented.

When we take the square and the cube of the number 69, we get two numbers that together use *all* the ten digits exactly once. $69^2 = 4,761$, and

$69^3 = 328,509$. Is there any other number for which this peculiarity is true?

Consider the following: $6,667^2 = 44,448,889$. Then multiply this result, 44,448,889, by 3 to get 133,346,667. We notice that the last four digits are the same as the four digits of the number we began with—6,667. Although this is a neat pattern, one that perhaps is more dramatic can be shown when we take the number 625 to any power. The resulting number will always end with the last three digits being 625.

There are only two such numbers of three digits that have this property. The other is 376, as the following shows.

$$625^k$$

625^1	=	**625**
625^2	=	390,**625**
625^3	=	244,140,**625**
625^4	=	152,587,890,**625**
625^5	=	95,367,431,640,**625**
625^6	=	59,604,644,775,390,**625**
625^7	=	37,252,902,984,619,140,**625**
625^8	=	23,283,064,365,386,962,890,**625**
625^9	=	14,551,915,228,366,851,806,640,**625**
625^{10}	=	9,094,947,017,729,282,379,150,390,**625**

. . .

$$376^k$$

376^1	=	**376**
376^2	=	141,**376**
376^3	=	53,157,**376**
376^4	=	19,987,173,**376**
376^5	=	7,515,177,189,**376**
376^6	=	2,825,706,623,205,**376**
376^7	=	1,062,465,690,325,221,**376**
376^8	=	399,487,099,562,283,237,**376**
376^9	=	150,207,149,435,418,497,253,**376**
376^{10}	=	56,477,888,187,717,354,967,269,**376**

. . .

If one questions whether there are two-digit numbers that have this property, the answer is clearly yes. We notice from above, they are 25 and 76.

Number relationships are boundless. For years some mathematicians have stumbled upon these while others have searched for these patterns. Some of them seem a bit far-fetched but nonetheless can be appealing to us from a recreational point of view. For example, consider taking any three-digit number that is multiplied by a five-digit number, all of whose digits are the same. The result will be a number where, when you add its last five digits to the remaining digits, a number will result in which all digits are the same. Here are a few such examples:

$237 \cdot 33,333 = 7,899,921$, then $78 + 99,921 = 99,999$.
$357 \cdot 77,777 = 27,766,389$, then $277 + 66,389 = 66,666$.
$789 \cdot 44,444 = 35,066,316$, then $350 + 66,316 = 66,666$.
$159 \cdot 88,888 = 14,133,192$, then $141 + 33,192 = 33,333$.

These amazing number peculiarities, although entertaining, allow us to exhibit the beauty of mathematics to win over those individuals who have not experienced seeing mathematics from this point of view. We offer some more of these here to further entice the reader.

SOME OTHER NUMBER CURIOSITIES

Here is a number equal to the sum of the fourth powers of its digits: $8,208 = 8^4 + 2^4 + 0^4 + 8^4$.

There are numbers where the square of the number is comprised of digits that can be partitioned into two squares, as with the following:

$$
\begin{array}{rcccl}
7^2 & = & 4\,\overline{9} & = & 2^2\,\overline{3^2} \\
13^2 & = & 16\,\overline{9} & = & 4^2\,\overline{3^2} \\
19^2 & = & 36\,\overline{1} & = & 6^2\,\overline{1^2} \\
35^2 & = & 1\,\overline{225} & = & 1^2\,\overline{15^2} \\
38^2 & = & 144\,\overline{4} & = & 12^2\,\overline{2^2} \\
57^2 & = & 324\,\overline{9} & = & 18^2\,\overline{3^2} \\
223^2 & = & 49\,\overline{729} & = & 7^2\,\overline{27^2}
\end{array}
$$

The smallest and largest square numbers containing the digits 1 to 9 are $11,826^2 = 139,854,276$, and $30,384^2 = 923,187,456$.

The smallest and largest square numbers containing the digits 0 to 9 are $32,043^2 = 1,026,753,849$, and $99,066^2 = 9,814,072,356$.

There are other patterns that we can discover, such as the following:

$$19 = 1 \cdot 9 + (1 + 9)$$
$$29 = 2 \cdot 9 + (2 + 9)$$
$$39 = 3 \cdot 9 + (3 + 9)$$
$$49 = 4 \cdot 9 + (4 + 9)$$
$$59 = 5 \cdot 9 + (5 + 9)$$
$$69 = 6 \cdot 9 + (6 + 9)$$
$$79 = 7 \cdot 9 + (7 + 9)$$
$$89 = 8 \cdot 9 + (8 + 9)$$
$$99 = 9 \cdot 9 + (9 + 9)$$

And another curious pattern is:

$$3^2 + 4^2 = 5^2 = 25$$
$$10^2 + 11^2 + 12^2 = 13^2 + 14^2 = 365$$
$$21^2 + 22^2 + 23^2 + 24^2 = 25^2 + 26^2 + 27^2 = 2,030$$
$$36^2 + 37^2 + 38^2 + 39^2 + 40^2 = 41^2 + 42^2 + 43^2 + 44^2 = 7,230$$
$$55^2 + 56^2 + 57^2 + 58^2 + 59^2 + 60^2 = 61^2 + 62^2 + 63^2 + 64^2 + 65^2 = 19,855$$

This can lead us to:

$$1 + 2 = 3$$
$$4 + 5 + 6 = 7 + 8$$
$$9 + 10 + 11 + 12 = 13 + 14 + 15$$
$$16 + 17 + 18 + 19 + 20 = 21 + 22 + 23 + 24$$
and so on.

One might ask if a number could be equal to the sum of the factorials[19] of each of the digits of that number. There are only four known numbers that satisfy that question. One such number is $145 = 1! + 4! + 5! = 1 + 24 + 120$. Aside from the two trivial cases, which are the numbers 1 and 2, the only other number that seems to fit that requirement is the number 40,585, since it is equal to $4! + 0! + 5! + 8! + 5! = 24 + 1 + 120 + 40,320 + 120$.

We can also form a nice pattern with the squares evolving from these sums:

$$1$$
$$1 + 1$$
$$1 + 2 + 1$$
$$1 + 2 + 3 + 2 + 1$$
$$1 + 2 + 3 + 4 + 3 + 2 + 1$$
$$1 + 2 + 3 + 4 + 5 + 4 + 3 + 2 + 1$$

This continues, giving us the sum of the numbers in each row as a perfect square.

There are numbers that can be found where the sum of all their divisors is a perfect square. Some of these numbers are provided in figure 1.8.

n	Sum of All Its Divisors				
3		$1 + 3$	$=$	4	$= 2^2$
22		$1 + 2 + 11 + 22$	$=$	36	$= 6^2$
66	$1 + 2 + 3 + 6 + 11 + 22 + 33 + 66$		$=$	144	$= 12^2$
70	$1 + 2 + 5 + 7 + 10 + 14 + 35 + 70$		$=$	144	$= 12^2$
81		$1 + 3 + 9 + 27 + 81$	$=$	121	$= 11^2$
94		$1 + 2 + 47 + 94$	$=$	144	$= 12^2$
115		$1 + 5 + 23 + 115$	$=$	144	$= 12^2$
119		$1 + 7 + 17 + 119$	$=$	144	$= 12^2$
170	$1 + 2 + 5 + 10 + 17 + 34 + 85 + 170$		$=$	324	$= 18^2$
. . .					

Figure 1.8

PRIME NUMBERS AND PRIMARY NUMBERS

Recall that *prime numbers* are natural numbers (we will use the symbol **N** to represent the set of natural numbers) that have exactly two divisors. In other words, a prime number is a number greater than 1 whose divisors are 1 and the number itself. Let's consider a set of natural numbers each of which when divided by 4 leaves a remainder of 1. Symbolically we can write that set as $M_{4n+1} = \{4n + 1 \mid n \in \mathbf{N}\} = \{1, 5, 9, 13, \ldots\}$. This could be read as the set of all natural numbers in the form of $4n + 1$.

If we inspect this a bit further, we will notice a curiosity when any of the two numbers in the set M_{4n+1} are multiplied.

Take for example the following pairs of these numbers:
$$5 \cdot 9 = 45 = 4 \cdot 11 + 1, \, 17 \cdot 29 = 493 = 4 \cdot 123 + 1.$$

It appears that each of these products will also end up being a member of the set M_{4n+1}, since when each is divided by 4, they leave a remainder of 1. When this happens we say that this set is *closed* under the operation of multiplication. That is to say that whenever we multiply two members of the set, the result is another member of the same set. Another example of a closed set from the set of natural numbers is the set of even numbers under the operation of multiplication, under subtraction, and under addition. With simple algebra numbers we can justify this generalization.

If we let $a = 4n_1 + 1$, and $b = 4n_2 + 1$, then the product is as follows, where n_1 and n_2 are any natural numbers: $a \cdot b = (4n_1 + 1)(4n_2 + 1) = 16n_1n_2 + 4n_1 + 4n_2 + 1 = 4(4n_1n_2 + n_1 + n_2) + 1 = 4n_3 + 1$, where $n_3 = 4n_1n_2 + n_1 + n_2 \in \mathbf{N}$. It then follows that $a \cdot b$ is also in M_{4n+1}.

Once again, let's recall that a prime number is one that has exactly two divisors: 1 and the number itself. By saying that it has exactly two divisors, we exclude the number 1 from the set of primes. For example, between 1 and 10 there are exactly four prime numbers, and between 1 and 100 there are twenty-five prime numbers, as shown here: 2, 3, 5, 7, 11, 13, 17, 19, 23, 29, 31, 37, 41, 43, 47, 53, 59, 61, 67, 71, 73, 79, 83, 89, 97.

Euclid (360–280 BCE) has already proven that there are an infinite number of primes.

Let us once again consider the set M_{4n+1}, where we will call the members of M_{4n+1} *primary numbers*, if they have exactly two divisors, 1 and the number itself. You'll notice the analogy to the prime numbers in the set from which they come, the natural numbers **N**.

Symbolically we would write this as $p \in M_{4n+1}$, and p is a primary number if it has exactly two factors, 1 and p, in the set. All of the other numbers in our set M_{4n+1} would then be considered *composite numbers*.

Let us now try to find which members of the set M_{4n+1} have exactly two divisors. Let's consider the following factorizations of the first few members of this set.

$(1 = 1)$	1 is not a primary number (it has not exactly two factors in the set M_{4n+1}, namely, only one, the 1).
$5 = 1 \cdot 5$	**5 is a primary number** (it has exactly two factors in the set M_{4n+1}, namely, only 1 and 5).
$9 = 3 \cdot 3$	**9 is a primary number** (it has exactly two factors in the set M_{4n+1}, namely, only 1 and 9; $3 \notin M_{4n+1}$).
$13 = 1 \cdot 13$	**13 is a primary number** (it has exactly two factors in the set M_{4n+1}, namely, only 1 and 13).
$17 = 1 \cdot 17$	**17 is a primary number** (it has exactly two factors in the set M_{4n+1}, namely, only 1 and 17).
$21 = 3 \cdot 7$	**21 is a primary number** (it has exactly two factors in the set M_{4n+1}, namely, only 1 and 21; $3 \notin M_{4n+1}$, $7 \notin M_{4n+1}$).
$25 = 5 \cdot 5$	25 is not a primary number (it has three factors in the set M_{4n+1}, namely, 1, 5, and 25). This is the first factorization with three factors.
$29 = 1 \cdot 29$	**29 is a primary number** (it has exactly two factors in the set M_{4n+1}, namely, only 1 and 29).
$33 = 3 \cdot 11$	**33 is a primary number** (it has exactly two factors in the set M_{4n+1}, namely, only 1 and 33; $3 \notin M_{4n+1}$, $11 \notin M_{4n+1}$).
$37 = 1 \cdot 37$	**37 is a primary number** (it has exactly two factors in the set M_{4n+1}, namely, only 1 and 37).
$41 = 1 \cdot 41$	**41 is a primary number** (it has exactly two factors in the set M_{4n+1}, namely, only 1 and 41).
$45 = 5 \cdot 9$	45 is not a primary number (it has four factors in the set M_{4n+1}, namely, 1, 5, 9, and 45). This is the first factorization with four factors.
$49 = 7 \cdot 7$	**49 is a primary number** (it has exactly two factors in the set M_{4n+1}, namely, only 1 and 39; $7 \notin M_{4n+1}$).
$53 = 1 \cdot 53$	**53 is a primary number** (it has exactly two factors in the set M_{4n+1}, namely, 1 and 53).
$57 = 3 \cdot 19$	**57 is a primary number** (it has exactly two factors in the set M_{4n+1}, namely, only 1 and 39; $3 \notin M_{4n+1}$, $19 \notin M_{4n+1}$).
$61 = 1 \cdot 61$	**61 is a primary number** (it has exactly two factors in the set M_{4n+1}, namely, 1 and 61).
$65 = 5 \cdot 13$	65 is not a primary number (it has four factors in the set M_{4n+1}, namely, 1, 5, 13, and 65).

and so on.

We then have the primary numbers in the set $M_{4n+1} = \{4n + 1 \mid n \in \mathbf{N}\}$ up to 65 as the numbers: 5, 9, 13, 17, 21, 29, 33, 37, 41, 49, 53, 57, 61.

Notice that the numbers 1, 25, 45, and 65 are not considered primary numbers, since they have a factorization in members of the set M_{4n+1}.

Curiously, we now have primary numbers (9, 21, 33, 49, and 57) that are not prime numbers, since in the set M_{4n+1} they have exactly two divisors, when we consider only the numbers in the set M_{4n+1} as possible divisors. Here is a list of the primary numbers between 65 and 200 in the set M_{4n+1}.

$$81 \,(= 9 \cdot 9)$$
$$85 \,(= 5 \cdot 17)$$
$$105 \,(= 5 \cdot 21)$$
$$117 \,(= 9 \cdot 13)$$
$$125 \,(= 5 \cdot 25)$$
$$145 \,(= 5 \cdot 29)$$
$$165 \,(= 5 \cdot 33)$$
$$169 \,(= 13 \cdot 13)$$
$$185 \,(= 5 \cdot 37)$$
$$189 \,(= 9 \cdot 21)$$

As we draw a comparison to the set of natural numbers, where we know that each member of the set can be factored uniquely in terms of prime-number members, we would expect that the members of the set M_{4n+1} can also be factored uniquely in terms of primary numbers. Curiously, this is not true. In order for us to show that something like this is not true, we merely have to show one example that contradicts this conjecture. Consider the number 325. Now, notice that we can represent it in two ways, where it is factored in terms of primary numbers: $325 = 5 \cdot 65 = 13 \cdot 25$. Yet this does not support our concern, since the numbers 5, 13, 25, and 65 include some composite (non-prime) numbers, and therefore these are not considered primary-number factorizations. However, we point this out to show that the number 325 can be factored more than one way. The same can be said for the following factorization of the number 405, where $405 = 5 \cdot 81 = 9 \cdot 45$, and only one factorization contains only primary numbers, since 45 is not a primary number.

The smallest number in the set M_{4n+1} that can be factored two ways in terms of primary numbers is the number $693 = 4 \cdot 173 + 1$. It can be factored as follows: $693 = 9 \cdot 77 = 21 \cdot 33$, where each of the factors is a primary number. This destroys the analogy to the natural numbers that we tried to set up earlier. Among the natural numbers, the prime numbers can be factored in exactly one way with two members of the set of naturals. In contrast, among the numbers in the set M_{4n+1}, the primary numbers need not always be factorable uniquely with two primary numbers, as we can see with the next larger example: $1{,}089 = 9 \cdot 121 = 33 \cdot 33$.

Perhaps with this curiosity, we will not take for granted from here on out that every member of the natural numbers can be expressed uniquely in terms of prime factors and respect it! We now have an illustration of a set (M_{4n+1}), where primary numbers are not necessarily uniquely factorable into two primary numbers.

An ambitious reader may wish to carry this investigation to other sets, such as M_{2n+1}, M_{3n+1}, M_{5n+1}, or M_{4n-1}, to investigate how the primary numbers of these sets behave with regard to their comparison to the prime numbers from the set of natural numbers.

PERFECT NUMBERS

You may be wondering now what makes a number perfect? Mathematicians define a number to be a *perfect number* when the sum of its factors (excluding the number itself) is equal to the number. For example, the smallest perfect number is 6, because the sum of its factors (of its proper positive divisors) is $1 + 2 + 3 = 6$. The next larger perfect number is 28, since again the sum of its factors is $1 + 2 + 4 + 7 + 14 = 28$. Perfect numbers have fascinated mathematicians for centuries. The ancient Greeks knew of the first four perfect numbers, and even Euclid established a formula for generating these perfect numbers. These first two perfect numbers were regarded by ancient biblical scholars as perfect in that the biblical creation was accomplished in six days and the lunar month was twenty-eight days. The next two perfect numbers (496 and 8,128) are attributed to

Nicomachus (ca. 100). Yet it was not until the eighteenth century that the Swiss mathematician Leonhard Euler (1707–1783) proved that a formula developed by Euclid would generate all even perfect numbers.

Let us now consider what Euclid developed as a generalized formula for generating additional perfect numbers. The formula for generating even perfect numbers is $2^{n-1}(2^n - 1)$, with the condition that $2^n - 1$ is a prime number, and is referred to as a *Mersenne prime number*,[20] M_p. For one thing, if n is a composite number, then $2^n - 1$ will surely be composite, thereby making such a value of n not one to generate a perfect number from Euclid's formula. You can demonstrate this with some very elementary algebra, as we will show here.

Suppose n is an even composite number, say $2x$; the expression $2^n - 1$ then becomes the difference of two squares, which is always factorable, and is, therefore, a composite number:

$$2^{2x} - 1 = \left(2^x - 1\right)\left(2^x + 1\right).$$

If n is an odd composite number, as in the expression $2^{pq} - 1$, it becomes a factorable term as

$$2^{pq} - 1 = \left(2^p\right)^q - 1 = \left(2^p - 1\right)\left(\left(2^p\right)^{q-1} + \left(2^p\right)^{q-2} + \left(2^p\right)^{q-3} + \ldots + \left(2^2\right) + 1\right).$$

This is not to say that whenever n is prime that $2^n - 1$ will also be prime. For example, when $n = 11$, we have $2^{11} - 1 = 2,048 - 1 = 2,047 = 23 \cdot 89$, and is therefore, not prime. Therefore, we have to be careful to make sure that n is a prime number and that it also generates $2^n - 1$ to be a prime.

There have been some bumps in the road that mathematicians encountered as they pursued the search for further perfect numbers. For example, the French mathematician Marin Mersenne (1588–1648), who studied perfect numbers, corrected the list of twenty-four perfect numbers, published by the seventeenth-century mathematician Peter Bungus in his book *Numerorun Mysteria* (1644), stating that only eight of these were correct (i.e., the first eight on the list in the table in figure 1.9). However, Mersenne offered to add three more numbers to this list of perfect numbers (namely, those where n in Euclid's formula had values 67, 127, and 257). By 1947, it was proved that only the number 127 was correct, and at that time two more perfect numbers with values of *n* being 89 and 107 were added to the list of perfect numbers.

Curiosities abound when it comes to searching for perfect numbers. On March 26, 1936, the *New York Herald Tribune* published an article that claimed that Dr. Samuel I. Krieger discovered a perfect number with more than nineteen digits.

In fact, his proposed number had 155 digits! He even presented this supposedly perfect number as

$2^{513} - 2^{256}$

$= 26,815,615,859,885,194,199,148,049,996,411,692,254,958,731,641,184,$
$786,755,447,122,887,443,528,060,146,978,161,514,511,280,138,383,284,$
$395,055,028,465,118,831,722,842,125,059,853,682,308,859,384,882,$
$528, 256.$

Although mathematicians previously dispelled this notion, it was not until 1952 that, with the aid of a computer, the number $2^{257}-1$ was shown to be a composite number, and, therefore, not capable of generating a perfect number. Mathematics journals chided the newspaper for its eagerness to report a story before verifying its accuracy.

Mathematicians are always fascinated with perfect numbers and consequently are always in search for further members of this set of perfect numbers.[21] As of this publication, there are forty-eight known perfect numbers, as shown in the table in figure 1.9.[22]

n	Perfect Number	Digits	Year Discovered
2	6	1	Greeks
3	28	2	Greeks
5	496	3	Greeks
7	8128	4	Greeks
13	33550336	8	1456
17	8589869056	10	1588
19	137438691328	12	1588
31	2305843008139952128	19	1772
61	2658455991569831744654692615953842176	37	1883
89	19156194260823610729479337808430363813099732 1548169216	54	1911
107	131640364...783728128	65	1914
127	144740111...199152128	77	1876
521	235627234...555646976	314	1952
607	141053783...537328128	366	1952
1279	541625262...984291328	770	1952
2203	108925835...453782528	1327	1952
2281	994970543...139915776	1373	1952
3217	335708321...628525056	1937	1957
4253	182017490...133377536	2561	1961
4423	407672717...912534528	2663	1961
9689	114347317...429577216	5834	1963
9941	598885496...073496576	5985	1963
11213	395961321...691086336	6751	1963
19937	931144559...271942656	12003	1971
21701	100656497...141605376	13066	1978
23209	811537765...941666816	13973	1979
44497	365093519...031827456	26790	1979
86243	144145836...360406528	51924	1982
110503	136204582...603862528	66530	1988
132049	131451295...774550016	79502	1983
216091	278327459...840880128	130100	1985
756839	151616570...565731328	455663	1992
859433	838488226...416167936	517430	1994
1257787	849732889...118704128	757263	1996
1398269	331882354...723375616	841842	1996
2976221	194276425...174462976	1791864	1997
3021377	811686848...022457856	1819050	1998
6972593	955176030...123572736	4197919	1999
13466917	427764159...863021056	8107892	2001
20996011	793508909...206896128	12640858	2003
24036583	448233026...572950528	14471465	2004
25964951	746209841...791088128	15632458	2005
30402457	497437765...164704256	18304103	2005
32582657	775946855...577120256	19616714	2006
37156667	204534225...074480128	22370543	2008
42643801	144285057...377253376	25674127	2009
43112609	500767156...145378816	25956377	2008
57885161	169296395...270130176	34850340	2013

Figure 1.9

There are lots of curious characteristics of perfect numbers. For example, we notice that they all end in either 28 or 6, and that is preceded by an odd digit. Mathematicians have been in search for odd perfect numbers, and as we can see, our list consists only of even perfect numbers. To date, they say, with confidence, that there are no odd perfect numbers less than $10^{1,500}$.

There are many unusual characteristics of perfect numbers beyond those that define them, namely that the sum of their divisors equals the number itself. Here are some curious characteristics of perfect numbers of the form $2^{n-1}(2^n - 1)$.

First, they are the sum of the first consecutive natural numbers as shown below:

$6 = 1 + 2 + 3$

$28 = 1 + 2 + 3 + 4 + 5 + 6 + 7$

$496 = 1 + 2 + 3 + 4 + 5 + 6 + 7 + 8 + 9 + \ldots + 29 + 30 + 31$

$8128 = 1 + 2 + 3 + 4 + 5 + 6 + 7 + 8 + 9 + \ldots + 125 + 126 + 127$

$33550336 = 1 + 2 + 3 + 4 + 5 + 6 + 7 + 8 + 9 + \ldots + 8189 + 8190 + 8191$,

and so on.

If this isn't enough, we also notice with the exception of the first perfect number, 6, they are equal to the sum of consecutive odd cubes, as shown below:

$28 = 1^3 + 3^3$

$496 = 1^3 + 3^3 + 5^3 + 7^3$

$8,128 = 1^3 + 3^3 + 5^3 + 7^3 + 9^3 + 11^3 + 13^3 + 15^3$

$33,550,336 = 1^3 + 3^3 + 5^3 + \ldots + 123^3 + 125^3 + 127^3$

Now, you may wonder, how can it be that suddenly perfect numbers are equal to the sum of the cubes of a sequence of odd numbers? This can be easily justified through some elementary algebra. We recall that the perfect number must take the form of $2^{n-1}(2^n - 1)$, where $2^n - 1$ is a prime number. We will take a moment here to show how we can justify that each perfect number is the sum of the first 2^k odd numbers, where $k = \frac{1}{2}(n-1)$, except for $n = 2$.

We should recall that the following are true:

$$S_1 = 1 + 2 + 3 + 4 + \cdots + q = \frac{q(q+1)}{2}$$

$$S_2 = 1^2 + 2^2 + 3^2 + 4^2 + \cdots + q^2 = \frac{q(q+1)(2q+1)}{6}$$

$$S_3 = 1^3 + 2^3 + 3^3 + 4^3 + \cdots + q^3 = \frac{q^2(q+1)^2}{4} = S_1^2$$

Let us now look at the sum of the cubes of the odd numbers. We can write that as follows:

$$S = 1^3 + 3^3 + 5^3 + 7^3 + \cdots + (2q-1)^3 = \sum_{i=1}^{q} (2i-1)^3.$$

With some algebraic manipulation we can show that this is equal to the following:

$$S = \sum_{i=1}^{q}(2i-1)^3 = \sum_{i=1}^{q}(8i^3 - 12i^2 + 6i - 1) = 8 \cdot \frac{q^2(q+1)^2}{4} - 12 \cdot \frac{q(q+1)(2q+1)}{6} + 6 \cdot \frac{q(q+1)}{2} - q = q^2(2q^2 - 1).$$

Notice how we inserted the first three formulas, which we generated above, into this last equation.

If we now let $q = 2^k$, then $S = 2^{2k}(2^{2k+1} - 1)$, then we notice that this would generate the perfect numbers $2^{n-1}(2^n - 1)$, when $n = 2k + 1$, which are odd numbers. This, then, shows how we go from the general formula for the perfect numbers to the sum of the cubes of the odd numbers. Although this may be a little bit of tough elementary algebra, we provided it here to show how we can justify some of these mathematical curiosities.

All perfect numbers must have an even number of divisors. If we take the reciprocals of the divisors of any perfect number (now including the number itself), their sum will always be equal to 2. This can be seen from the first few perfect numbers, as follows:

$$\frac{1}{1}+\frac{1}{2}+\frac{1}{3}+\frac{1}{6}=2$$

$$\frac{1}{1}+\frac{1}{2}+\frac{1}{4}+\frac{1}{7}+\frac{1}{14}+\frac{1}{28}=2$$

$$\frac{1}{1}+\frac{1}{2}+\frac{1}{4}+\frac{1}{8}+\frac{1}{16}+\frac{1}{31}+\frac{1}{62}+\frac{1}{124}+\frac{1}{248}+\frac{1}{496}=2$$

MORE CURIOUS NUMBER PATTERNS

Number patterns have fascinated folks for centuries. They can be discovered by children or adults—or by mathematicians while engaged in an unrelated research project. For example, if we consider the fraction $\frac{1}{81}$ and take its decimal equivalent—we get a rather nice result: 0.**012345679** 012345679 **012345679** . . . = $0.\overline{012345679}$.

Surprisingly, we notice that the digits (without the 8) appear in order—continuously!

One might also stumble upon a multiplication example that uses all the nine non-zero digits exactly once—both in the multiplicand and in the product. Here are some examples of this phenomenon:

$$81{,}274{,}365 \cdot 9 = 731{,}469{,}285$$
$$72{,}645{,}831 \cdot 9 = 653{,}812{,}479$$
$$58{,}132{,}764 \cdot 9 = 523{,}194{,}876$$
$$76{,}125{,}483 \cdot 9 = 685{,}129{,}347$$

There are times that we can discover such simple relationships as the following:

$$
\begin{aligned}
1 &= 1 & &= 1 \cdot 1 = 1^2 \\
1+2+1 &= 2+2 & &= 2 \cdot 2 = 2^2 \\
1+2+3+2+1 &= 3+3+3 & &= 3 \cdot 3 = 3^2 \\
1+2+3+4+3+2+1 &= 4+4+4+4 & &= 4 \cdot 4 = 4^2 \\
1+2+3+4+5+4+3+2+1 &= 5+5+5+5+5 & &= 5 \cdot 5 = 5^2 \\
1+2+3+4+5+6+5+4+3+2+1 &= 6+6+6+6+6+6 & &= 6 \cdot 6 = 6^2 \\
1+2+3+4+5+6+7+6+5+4+3+2+1 &= 7+7+7+7+7+7+7 & &= 7 \cdot 7 = 7^2 \\
1+2+3+4+5+6+7+8+7+6+5+4+3+2+1 &= 8+8+8+8+8+8+8+8 & &= 8 \cdot 8 = 8^2 \\
1+2+3+4+5+6+7+8+9+8+7+6+5+4+3+2+1 &= 9+9+9+9+9+9+9+9+9 & &= 9 \cdot 9 = 9^2
\end{aligned}
$$

Once we recall that the product of 37 and 3 is such a nice number as 111, we can begin to search for a further pattern—such as considering multiples of 3. Then extending this to the sum of the digits of each of the products gives us further patterns—namely that the sum of the digits of any multiple of 3 is also a multiple of 3.

Number patterns often evolve through some curious calculation results, as we will see here. One of the most visually appealing arithmetic phenomena can be seen in the following "oddities." One often asks, why do such unusual number patterns exist? Sometimes the best answer is that it is a peculiarity of the base-ten system. Although this may not satisfy everyone, for many it may suffice. Therefore, we say to you, the reader, just enjoy and marvel at these patterns. We begin with a pattern that we have presented earlier:

$$
\begin{aligned}
1 \cdot 1 &= 1 \\
11 \cdot 11 &= 121 \\
111 \cdot 111 &= 12321 \\
1111 \cdot 1111 &= 1234321 \\
11111 \cdot 11111 &= 123454321 \\
111111 \cdot 111111 &= 12345654321 \\
1111111 \cdot 1111111 &= 1234567654321 \\
11111111 \cdot 11111111 &= 123456787654321 \\
111111111 \cdot 111111111 &= 12345678987654321
\end{aligned}
$$

Another number pattern appears when we replace the 1s with 9s to get the following:

$$9 \cdot 9 = 81$$
$$99 \cdot 99 = 9801$$
$$999 \cdot 999 = 998001$$
$$9999 \cdot 9999 = 99980001$$
$$99999 \cdot 99999 = 9999800001$$
$$999999 \cdot 999999 = 999998000001$$
$$9999999 \cdot 9999999 = 99999980000001$$

and so on.

With 9s we can still produce another esthetically pleasing pattern as shown below.

$$999{,}999 \cdot \quad 1 = \mathbf{0{,}999{,}999}$$
$$999{,}999 \cdot \quad 2 = \mathbf{1{,}999{,}998}$$
$$999{,}999 \cdot \quad 3 = \mathbf{2{,}999{,}997}$$
$$999{,}999 \cdot \quad 4 = \mathbf{3{,}999{,}996}$$
$$999{,}999 \cdot \quad 5 = \mathbf{4{,}999{,}995}$$
$$999{,}999 \cdot \quad 6 = \mathbf{5{,}999{,}994}$$
$$999{,}999 \cdot \quad 7 = \mathbf{6{,}999{,}993}$$
$$999{,}999 \cdot \quad 8 = \mathbf{7{,}999{,}992}$$
$$999{,}999 \cdot \quad 9 = \mathbf{8{,}999{,}991}$$
$$999{,}999 \cdot 10 = \mathbf{9{,}999{,}990}$$

Here is another number pattern, where the number 9 is multiplied by a number representing the consecutive natural numbers—increasing by 1 each time—and then added to the initial natural numbers consecutively.

$$0 \cdot 9 + 1 = 1$$
$$1 \cdot 9 + 2 = 11$$
$$12 \cdot 9 + 3 = 111$$
$$123 \cdot 9 + 4 = 1111$$
$$1234 \cdot 9 + 5 = 11111$$
$$12345 \cdot 9 + 6 = 111111$$
$$123456 \cdot 9 + 7 = 1111111$$
$$1234567 \cdot 9 + 8 = 11111111$$
$$12345678 \cdot 9 + 9 = 111111111$$
$$123456789 \cdot 9 + 10 = 1111111111$$

We can consider number patterns that can be generated in a similar fashion, yet somewhat in the reverse of the previous one. However, this time the generated numbers consist of all 8s.

$$0 \cdot 9 + 8 = 8$$
$$9 \cdot 9 + 7 = 88$$
$$98 \cdot 9 + 6 = 888$$
$$987 \cdot 9 + 5 = 8888$$
$$9876 \cdot 9 + 4 = 88888$$
$$98765 \cdot 9 + 3 = 888888$$
$$987654 \cdot 9 + 2 = 8888888$$
$$9876543 \cdot 9 + 1 = 88888888$$
$$98765432 \cdot 9 + 0 = 888888888$$

Now that we have introduced the 8s in a rather dramatic fashion, we shall use them as the multiplier with numbers consisting of increasing natural numbers, and each time adding successive natural numbers. Appreciating the number pattern shown here is more pleasing than trying to explain this phenomenon, which could conceivably detract from its beauty.

$$1 \cdot 8 + 1 = 9$$
$$12 \cdot 8 + 2 = 98$$
$$123 \cdot 8 + 3 = 987$$
$$1234 \cdot 8 + 4 = 9876$$
$$12345 \cdot 8 + 5 = 98765$$
$$123456 \cdot 8 + 6 = 987654$$
$$1234567 \cdot 8 + 7 = 9876543$$
$$12345678 \cdot 8 + 8 = 98765432$$
$$123456789 \cdot 8 + 9 = 987654321$$

This time, using 1s and 8s, we come up with another number pattern that we can appreciate.

$1 \cdot 8$	=	8
$11 \cdot 88$	=	968
$111 \cdot 888$	=	98,568
$1,111 \cdot 8,888$	=	9,874,568
$11,111 \cdot 88,888$	=	987,634,568
$111,111 \cdot 888,888$	=	98,765,234,568
$1,111,111 \cdot 8,888,888$	=	9,876,541,234,568
$11,111,111 \cdot 88,888,888$	=	987,654,301,234,568
$111,111,111 \cdot 888,888,888$	=	98,765,431,901,234,568
$1,111,111,111 \cdot 8,888,888,888$	=	9,876,543,207,901,234,568

Proceeding along the number patterns, we offer the following patterns for consideration and appreciation.

It is rather strange that when we consider the decimal expansion of the fraction $\frac{1}{7} = 0.\textbf{142857}\,142857\,\textbf{142857}\ldots = 0.\overline{142857}$, we notice the pattern that evolves in the following tables.

$$1 \cdot 7 + 3 = 10$$
$$14 \cdot 7 + 2 = 100$$
$$142 \cdot 7 + 6 = 1000$$
$$1428 \cdot 7 + 4 = 10000$$
$$14285 \cdot 7 + 5 = 100000$$
$$142857 \cdot 7 + 1 = 1000000$$
$$1428571 \cdot 7 + 3 = 10000000$$
$$14285714 \cdot 7 + 2 = 100000000$$
$$142857142 \cdot 7 + 6 = 1000000000$$
$$1428571428 \cdot 7 + 4 = 10000000000$$
$$14285714285 \cdot 7 + 5 = 100000000000$$
$$142857142857 \cdot 7 + 1 = 1000000000000$$

For entertainment, we offer the following curious patterns.

$$7 \cdot 7 = 49$$
$$67 \cdot 67 = 4489$$
$$667 \cdot 667 = 444889$$
$$6667 \cdot 6667 = 44448889$$
$$66667 \cdot 66667 = 4444488889$$
$$666667 \cdot 666667 = 444444888889$$
$$6666667 \cdot 6666667 = 44444448888889$$
$$66666667 \cdot 66666667 = 4444444488888889$$
$$666666667 \cdot 666666667 = 444444444888888889$$

$$7 \cdot 9 = 63$$
$$77 \cdot 99 = 7623$$
$$777 \cdot 999 = 776223$$
$$7777 \cdot 9999 = 77762223$$
$$77777 \cdot 99999 = 7777622223$$
$$777777 \cdot 999999 = 777776222223$$

and so on.

The 7s do not hold a monopoly here; notice what occurs with the 4s.

$$4 \cdot 4 = 16$$
$$34 \cdot 34 = 1156$$
$$334 \cdot 334 = 111556$$
$$3334 \cdot 3334 = 11115556$$
$$33334 \cdot 33334 = 1111155556$$
$$333334 \cdot 333334 = 111111555556$$

Setting up patterns to search for relationships is essentially unlimited. Below, you will find a rather-strange multiplication of a palindromic number consisting of the initial natural numbers multiplied by the sum of its digits and leading to a rather-unusual result. Readers should continue to be encouraged to discover new patterns resulting from some form of systematic arithmetic.

$$1 \cdot 1 = 1^2 = 1$$
$$121 \cdot (1 + 2 + 1) = 22^2 = 484$$
$$12321 \cdot (1 + 2 + 3 + 2 + 1) = 333^2 = 110889$$
$$1234321 \cdot (1 + 2 + 3 + 4 + 3 + 2 + 1) = 4444^2 = 19749136$$
$$123454321 \cdot (1 + 2 + 3 + 4 + 5 + 4 + 3 + 2 + 1) = 55555^2 = 3086358025$$

Number curiosities continue with the number 76,923, which offers us some unusual peculiarities. However, before inspecting this peculiarity, we should notice that the decimal expansion of the fraction $\frac{1}{13}$ = 0.**076923** 076923 **076923** . . . = $0.\overline{076923}$. There arises the number 76,923, albeit a bit camouflaged. When 76,923 is multiplied by the numbers: 1, 10, 9, 12, 3, and 4, in succession, the products show an unexpected pattern. The digits of the products rotate around, as you can see in the list below. Notice how the first digit of the first product appears at the end of the second product. Then the first digit of the second product appears at the end of the third product, and so the pattern continues.

$$76{,}923 \cdot \ \ 1 = 076{,}923$$
$$76{,}923 \cdot 10 = 769{,}230$$
$$76{,}923 \cdot \ \ 9 = 692{,}307$$
$$76{,}923 \cdot 12 = 923{,}076$$
$$76{,}923 \cdot \ \ 3 = 230{,}769$$
$$76{,}923 \cdot \ \ 4 = 307{,}692$$

When we now take the same number, 76,923, and multiply it by the remaining initial natural numbers up to 12, we once again see a similar pattern emerging. However, when we multiply this number (76,923) by the next natural number, 13, we get the unexpected product 999,999.

$$76{,}923 \cdot \ \ 2 = 153{,}846$$
$$76{,}923 \cdot \ \ 7 = 538{,}461$$
$$76{,}923 \cdot \ \ 5 = 384{,}615$$
$$76{,}923 \cdot 11 = 846{,}153$$
$$76{,}923 \cdot \ \ 6 = 461{,}538$$
$$76{,}923 \cdot \ \ 8 = 615{,}384$$

When the number 76,923, is multiplied by 14, 15, 16, 17, and so on, we get a strange pattern, which compares to the products of this number when we multiplied it by the natural numbers in order (above). An ambitious reader might want to go beyond these consecutive multiplications to find other unusual patterns.

$76{,}923 \cdot 14 = 1\ 076922$	(compare to the multiple by 1)
$76{,}923 \cdot 15 = 1\ 153845$	(compare to the multiple by 2)
$76{,}923 \cdot 16 = 1\ 230768$	(compare to the multiple by 3)
$76{,}923 \cdot 17 = 1\ 307691$	(compare to the multiple by 4)
$76{,}923 \cdot 18 = 1\ 384614$	(compare to the multiple by 5)
$76{,}923 \cdot 19 = 1\ 461537$	(compare to the multiple by 6)
$76{,}923 \cdot 20 = 1\ 538460$	(compare to the multiple by 7)
$76{,}923 \cdot 21 = 1\ 615383$	(compare to the multiple by 8)
$76{,}923 \cdot 22 = 1\ 692306$	(compare to the multiple by 9)

$76,923 \cdot 23 = 1\ 769229$ (compare to the multiple by 10)
$76,923 \cdot 24 = 1\ 846152$ (compare to the multiple by 11)
$76,923 \cdot 25 = 1\ 923075$ (compare to the multiple by 12)
$76,923 \cdot 26 = 1\ 999998$ (compare to the multiple by 13)
$76,923 \cdot 27 = 2\ 076921$ (compare to the multiple by 14)
$76,923 \cdot 28 = 2\ 153844$ (compare to the multiple by 15)
$76,923 \cdot 29 = 2\ 230767$ (compare to the multiple by 16)

And the pattern continues indefinitely!

Let us now consider the number 142,857. First we notice that all the same digits from the original number, 142,857, are used for each of the products from 1 to 6.

A CYCLIC NUMBER LOOP

Take any integer from 1 to 6 and multiply it by 999,999 and then divide it by 7. You will get a number made up of the digits 1, 4, 2, 8, 5, 7. Not only that, but they will be in this order, yet starting from a different digit each time. This is the phenomenon of a *cyclic number*, which is a number of n digits that, when multiplied by each of the numbers $1, 2, 3, 4, \ldots, n$, produces a number that uses the same digits as the original number, but in a different order each time.

These numbers are called *Phoenix numbers*—after the bird that according to an ancient Egyptian legend rises youthfully from its ashes whenever it is burned and disappears. (See figure 1.10.)

142,857 · 1 = 142,857
142,857 · 2 = 285,714
142,857 · 3 = 428,571
142,857 · 4 = 571,428
142,857 · 5 = 714,285
142,857 · 6 = 857,142
142,857 · 7 = **999,999**

Figure 1.10

If we multiply the number 142,857 by 7, we get 999,999. Another peculiarity occurs when we multiply the number 142,857 by 8; we get 1,142,856. With some imagination, we could take the first digit and add it to the last digit, and we would be back to where we started with the number 142,857. Taking this a step further, when we multiply 142,857 by 9, we get 1,285,713. Using the same awkward technique by taking the first digit and adding it to the last digit, we get the product of the number 142,857 when multiplied by 2.

There is still more we can show with this unusual number. Clearly the sum of the digits for each of the products 1 to 6 is 27, which is one-half of the sum of the digits of the product when multiplied by 9. Furthermore, not only is the sum of the digits of each of the products 27, but if we take the sum of the digits vertically, we will also get a sum for each place value to be 27.

$$
\begin{array}{cccccc}
1 & 4 & 2 & 8 & 5 & 7 = 27 \\
2 & 8 & 5 & 7 & 1 & 4 = 27 \\
4 & 2 & 8 & 5 & 7 & 1 = 27 \\
5 & 7 & 1 & 4 & 2 & 8 = 27 \\
7 & 1 & 4 & 2 & 8 & 5 = 27 \\
8 & 5 & 7 & 1 & 4 & 2 = 27 \\
\hline
\end{array}
$$

27 27 27 27 27 27

We now come to a very peculiar aspect of this curious number. Suppose we multiply this number, 142,857, by another large number, say, 32,789,563,521, we get the product 4,684,218,675,919,497. We will break up this product in groups of six, beginning at the right side of the number, and then add the numbers.

$$
\begin{array}{r}
919,497 \\
218,675 \\
\underline{4,684} \\
1,142,856
\end{array}
$$

We are almost finished with this demonstration. All we now need to do is to take this "six-grouping" one step further: we will take the first 1 and add it to the remaining number to get the following:

$$
\begin{array}{r}
142,856 \\
\underline{1} \\
142,857
\end{array}
$$

By now you are probably not surprised to see that the number 142,857 appears again. Just to demonstrate this as not being "rigged," we will do this with another product of 142,857 and the large number 89,651,273,582,410,598 to get 12,807,311,990,162,430,798,486. Then, breaking this number up into groups of six, beginning on the right side, we get the following:

$$
\begin{array}{r}
798,486 \\
162,430 \\
311,990 \\
\underline{12,807} \\
1,285,713
\end{array}
$$

Once again, will take the first 1 and added to the number, and we notice the almost-anticipated result.

$$\begin{array}{r} 285{,}713 \\ 1 \\ \hline 285{,}714 = 142{,}857 \cdot 2 \end{array}$$

You may want to try other such products of 142,857 to verify this phenomenon.

BABYLONIAN MULTIPLICATION

The Babylonians had a rather peculiar way of multiplying numbers. They had developed a chart of numbers along with their squares, and with that they set up a rather-interesting way to multiply two randomly selected numbers. They were able to add and subtract numbers. If they chose to multiply two numbers, say, a and b, they first found the sum $(a + b)$ and then the difference $(a - b)$ of these two numbers. They then subtracted these two values and divided by four to get the originally required product. For example, consider the following. Suppose we wish to multiply 53 by 47. The process looked like this:

$$53 \cdot 47 \quad \frac{(53+47)^2-(53-47)^2}{4} = \frac{100^2-6^2}{4} = \frac{10000-36}{4} = \frac{9964}{4} = 2{,}491.$$

To consider the general case we would let $a = 53$ and $b = 47$, and then we have the following:

$$\frac{(a+b)^2-(a-b)^2}{4} = \frac{a^2+2ab+b^2-a^2+2ab-b^2}{4} = \frac{4ab}{4} = ab.$$

To what extent they knew this sophisticated algebra is probably open to some speculation. However, it did work, and they used it to multiply numbers.

The beauty in this is that we can also justify this procedure geometrically—however we will rely only on positive numbers. In figure 1.11, we show the large square side length $(a + b)^2$, and the smaller square side length $(a - b)^2$. Each of the rectangles has an area equal to $a \cdot b$.

Therefore, the difference between the areas of the two squares results in the four rectangles, one quarter of which, $\frac{(a+b)^2-(a-b)^2}{4}$, is the area of one rectangle, namely, $a \cdot b$, which is the desired product.

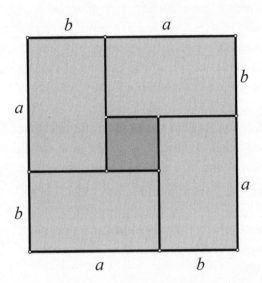

Figure 1.11

A further example of how this Babylonian method of multiplication can work for $a = 2.3$ and $b = 1.6$ is shown below.

$$2.3 \cdot 1.6 = \frac{(2.3+1.6)^2 - (2.3-1.6)^2}{4} = \frac{3.9^2 - 0.7^2}{4} = \frac{15.21 - 0.49}{4} = \frac{14.72}{4} = 3.68.$$

Remember, this clever technique was based on the fact that the Babylonians had a table of squares at their disposal. We are not trying to demonstrate a lack of need for a calculator, but rather to show how clever previous cultures that did not have our modern technology were.

RUSSIAN PEASANT'S METHOD OF MULTIPLICATION

There is a rather-curious way to do multiplication. It is said that the Russian peasants used a rather-strange, perhaps even primitive, method to multiply two numbers. It is actually quite simple, yet somewhat cumbersome. Let's take a look at the problem of finding the product of 73 · 82. We shall work this multiplication together. We begin by setting up a chart of two columns with the two members of the multiplication, 73 and 82, in the first row. (See figure 1.12.) One column will be formed by doubling each number to get the next, while the other column will take half the number and drop the remainder. Let us agree that our first column will be the doubling column, and the second column will be the halving column. Notice that by halving the odd number such as 41 (the second number in the second column) we get 20 with a remainder of 1 and we simply drop the 1. The rest of this halving process should now be clear.

Priority of Halving Numbers	Doubling	Halving
Even	73	82
Odd	**146**	**41**
Even	292	20
Even	584	10
Odd	**1168**	**5**
Even	2336	2
Odd	**4672**	**1**

Figure 1.12

In figure 1.12, we find the odd numbers in the halving column (here, the right-hand column), then get the sum of their partner numbers in the doubling column (in this case, the left column). These are highlighted in bold type. This sum gives us the originally required product of 73 and 82. In other words, 73 · 82 = 146 + 1,168 + 4,672 = 5,986.

In the example above, we chose to have the first column be the doubling column and the second column be the halving column. We could also have done this by halving the numbers in the first column and doubling those in the second. (See figure 1.13.)

Parity of Halving Numbers	Halving	Doubling
Odd	**73**	**82**
Even	36	164
Even	18	328
Odd	**9**	**656**
Even	4	1312
Even	2	2624
Odd	**1**	**5248**

Figure 1.13

To complete the multiplication, we find the odd numbers in the halving column and then get the sum of their partner numbers in the second column (now the doubling column). This gives us $73 \cdot 82 = 82 + 656 + 5{,}248 = 5{,}986$.

It is curious to see an algorithm that is so simple and, surprisingly, delivers the right answer. Yet in our age of technology such a situation is relegated to recreational mathematics—merely for entertainment.

By analyzing each step systematically, we can see how this algorithm works. (The asterisk indicates an odd number.)

$$*73 -- 82 \quad = (36 \cdot 2 + 1) \quad \cdot 82 \quad = 36\ 164 \quad + \mathbf{82} \quad = 5986$$
$$36 -- 164 \quad = (18 \cdot 2 + 0) \quad \cdot 164 \quad = 18 \cdot 328 \quad + 0 \quad = 5904$$
$$18 -- 328 \quad = (9 \cdot 2 + 0) \quad \cdot 328 \quad = 9 \cdot 656 \quad + 0 \quad = 5904$$
$$*9 -- 656 \quad = (4 \cdot 2 + 1) \quad \cdot 656 \quad = 4 \cdot 1312 \quad + \mathbf{656} \quad = 5904$$
$$4 -- 1312 \quad = (2 \cdot 2 + 0) \quad \cdot 1312 \quad = 2 \cdot 2624 \quad + 0 \quad = 5248$$
$$2 -- 2624 \quad = (1 \cdot 2 + 0) \quad \cdot 2624 = \quad 5248 \quad + 0 \quad = 5248$$
$$*1 -- 5248 \quad = (0 \cdot 2 + 1) \quad \cdot 5248 = \quad 0 \quad + \underline{\mathbf{5248}} = 5248$$

$$5986$$

For those familiar with the binary system (i.e., base 2), one can also explain this method with the following representation:

$$73 \cdot 82 = (1 \cdot 2^6 + 0 \cdot 2^5 + 0 \cdot 2^4 + 1 \cdot 2^3 + 0 \cdot 2^2 + 0 \cdot 2^1 + 1 \cdot 2^0) \cdot 82$$
$$= 2^0 \cdot 82 + 2^3 \cdot 82 + 2^6 \cdot 82$$
$$= 82 + 656 + 5248$$
$$= 5986$$

Whether or not you have a full understanding of the discussion of this method of multiplication, you should at least now have a deeper appreciation for the multiplication algorithm you learned in school, even though most people today multiply with a calculator. There are many other multiplication algorithms, yet the one shown here is perhaps one of the strangest, and it is through this strangeness that we can appreciate the powerful consistency of mathematics that allows us to conjure up such an algorithm.

SOME PRIME DENOMINATORS

Earlier we considered prime numbers, those that have no factors other than themselves and 1, but now we will consider fractions whose denominators are prime numbers (excluding 2 and 5) and where the decimal expansion yields an even number of repetitions. These will give us further material to discover patterns to amaze us.

We will now treat each one of these repeating periods as a number, and split this even-digit sequence of digits into two parts, and then add them. An amazing result appears.[23] In figure 1.14, we show the unexpected result of numbers emerging with all nines!

$$\frac{1}{7} = 0.\overline{142857}$$

142
857
999

$$\frac{1}{11} = 0.\overline{09}$$

0
9
9

$$\frac{1}{13} = 0.\overline{076923}$$

076
923
999

$$\frac{1}{17} = 0.\overline{0588235294117647}$$

05882352
94117647
99999999

$$\frac{1}{19} = 0.\overline{052631578947368421}$$

052631578
947368421
999999999

$$\frac{1}{23} = 0.\overline{0434782608695652173913}$$

04347826086
95652173913
99999999999

Figure 1.14

By the way, speaking of prime numbers, the numbers 139 and 149 are the first two prime numbers that differ by 10.

SOME MORE CURIOUS NUMBERS

We will now embark on a most-unusual numerical phenomenon. This time we will begin by considering unit fractions whose denominators are a multiple of 9, and not a multiple of 2 or of 5. Some examples of this would be the following fractions: $\frac{1}{27}, \frac{1}{63}, \frac{1}{81}, \frac{1}{99}, \frac{1}{117}, \frac{1}{153}$, and $\frac{1}{171}$. When we convert each of these fractions to decimals form—by dividing the denominator into the numerator—we get the following decimals:

$$\frac{1}{9\cdot3}=\frac{1}{27}=0.037037037037037037037037037037037\ldots=0.\overline{037}$$

$$\frac{1}{9\cdot7}=\frac{1}{63}=0.0158730158730158730158730015873\ldots=0.\overline{015873}$$

$$\frac{1}{9\cdot9}=\frac{1}{81}=0.012345679012345679012345679\ldots=0.\overline{012345679}$$

$$\frac{1}{9\cdot13}=\frac{1}{117}=0.0085470085470085470085470008547\ldots=0.\overline{008547}$$

$$\frac{1}{9\cdot17}=\frac{1}{153}=0.0065359477124183006535947712418\ldots=0.\overline{0065359477124183}$$

$$\frac{1}{9\cdot19}=\frac{1}{171}=0.0058479532163742690058479532163742\ldots=0.\overline{005847953216374269}$$

Thus far, nothing spectacular has really appeared with this decimal conversion of the unit fractions. What follows is truly an amazing curiosity of our number system. Follow along as we will now take the repeating digits (without the initial zeros) and multiply this newly formed number by the multiple of 9 that was used to get these denominators and observe the marvelous pattern that emerges. For example, in the first case we will multiply 37 by multiples of 3, since the denominator 27 was arrived at by multiplying 9 times 3. Therefore, we will multiply 37 by 3, by 6, by 9, and so on, until we reach 27.

$$37 \cdot 3 = 111$$
$$37 \cdot 6 = 222$$
$$37 \cdot 9 = 333$$
$$37 \cdot 12 = 444$$
$$37 \cdot 15 = 555$$
$$37 \cdot 18 = 666$$
$$37 \cdot 21 = 777$$
$$37 \cdot 24 = 888$$
$$37 \cdot 27 = 999$$

The next fraction's denominators were attained by taking 9 · 7. Now using multiples of 7 and multiplying them by the repeating part, 15,873, once again, we obtain an easily recognized pattern.

$$15\ 873 \cdot \ \ 7 = 111\ 111$$
$$15\ 873 \cdot 14 = 222\ 222$$
$$15\ 873 \cdot 21 = 333\ 333$$
$$15\ 873 \cdot 28 = 444\ 444$$
$$15\ 873 \cdot 35 = 555\ 555$$
$$15\ 873 \cdot 42 = 666\ 666$$
$$15\ 873 \cdot 49 = 777\ 777$$
$$15\ 873 \cdot 56 = 888\ 888$$
$$15\ 873 \cdot 63 = 999\ 999$$

Continuing along, we find these results each time delivering these amazing—yet pretty—numbers for the succeeding fractions listed earlier.

$$12345679 \cdot \ \ 9 = 111111111$$
$$12345679 \cdot 18 = 222222222$$
$$12345679 \cdot 27 = 333333333$$
$$12345679 \cdot 36 = 444444444$$
$$12345679 \cdot 45 = 555555555$$
$$12345679 \cdot 54 = 666666666$$
$$12345679 \cdot 63 = 777777777$$
$$12345679 \cdot 72 = 888888888$$
$$12345679 \cdot 81 = 999999999$$

$$8\,547 \cdot \ \ \ 13 = 111\ 111$$
$$8\,547 \cdot \ \ \ 26 = 222\ 222$$
$$8\,547 \cdot \ \ \ 39 = 333\ 333$$
$$8\,547 \cdot \ \ \ 52 = 444\ 444$$
$$8\,547 \cdot \ \ \ 65 = 555\ 555$$
$$8\,547 \cdot \ \ \ 78 = 666\ 666$$
$$8\,547 \cdot \ \ \ 91 = 777\ 777$$
$$8\,547 \cdot 104 = 888\ 888$$
$$8\,547 \cdot 117 = 999\ 999$$

$$65\ 359\ 477\ 124\ 183 \cdot 17 = 1\ 111\ 111\ 111\ 111\ 111\ 111$$
$$65\ 359\ 477\ 124\ 183 \cdot 34 = 2\ 222\ 222\ 222\ 222\ 222$$
$$65\ 359\ 477\ 124\ 183 \cdot 51 = 3\ 333\ 333\ 333\ 333\ 333$$
$$65\ 359\ 477\ 124\ 183 \cdot 68 = 4\ 444\ 444\ 444\ 444\ 444$$
$$65\ 359\ 477\ 124\ 183 \cdot 85 = 5\ 555\ 555\ 555\ 555\ 555$$
$$65\ 359\ 477\ 124\ 183 \cdot 102 = 6\ 666\ 666\ 666\ 666\ 666$$
$$65\ 359\ 477\ 124\ 183 \cdot 119 = 7\ 777\ 777\ 777\ 777\ 777$$
$$65\ 359\ 477\ 124\ 183 \cdot 136 = 8\ 888\ 888\ 888\ 888\ 888$$
$$65\ 359\ 477\ 124\ 183 \cdot 153 = 9\ 999\ 999\ 999\ 999\ 999$$

$$5\ 847\ 953\ 216\ 374\ 269 \cdot 19 = 111\ 111\ 111\ 111\ 111\ 111$$
$$5\ 847\ 953\ 216\ 374\ 269 \cdot 38 = 222\ 222\ 222\ 222\ 222\ 222$$
$$5\ 847\ 953\ 216\ 374\ 269 \cdot 57 = 333\ 333\ 333\ 333\ 333\ 333$$
$$5\ 847\ 953\ 216\ 374\ 269 \cdot 76 = 444\ 444\ 444\ 444\ 444\ 444$$
$$5\ 847\ 953\ 216\ 374\ 269 \cdot 95 = 555\ 555\ 555\ 555\ 555\ 555$$
$$5\ 847\ 953\ 216\ 374\ 269 \cdot 114 = 666\ 666\ 666\ 666\ 666\ 666$$
$$5\ 847\ 953\ 216\ 374\ 269 \cdot 133 = 777\ 777\ 777\ 777\ 777\ 777$$
$$5\ 847\ 953\ 216\ 374\ 269 \cdot 152 = 888\ 888\ 888\ 888\ 888\ 888$$
$$5\ 847\ 953\ 216\ 374\ 269 \cdot 171 = 999\ 999\ 999\ 999\ 999\ 999$$

You may want to extend this list of fractions and verify that the pattern continues.

Just to show further peculiarities, notice what happens when we reverse the order of one of our previous multiplications (the one evolving from $\frac{1}{81}$). Again, we get an unexpected, rather-curious pattern!

$$987654321 \cdot 9 = \mathbf{8}\ 88888888\ \mathbf{9}$$
$$987654321 \cdot 18 = \mathbf{1\ 7}\ 777777777\ \mathbf{8}$$
$$987654321 \cdot 27 = \mathbf{2\ 6}\ 66666666\ \mathbf{7}$$
$$987654321 \cdot 36 = \mathbf{3\ 5}\ 55555555\ \mathbf{6}$$
$$987654321 \cdot 45 = \mathbf{4\ 4}\ 44444444\ \mathbf{5}$$
$$987654321 \cdot 54 = \mathbf{5\ 3}\ 33333333\ \mathbf{4}$$
$$987654321 \cdot 63 = \mathbf{6\ 2}\ 22222222\ \mathbf{3}$$
$$987654321 \cdot 72 = \mathbf{7\ 1}\ 11111111\ \mathbf{2}$$
$$987654321 \cdot 81 = \mathbf{8\ 0}\ 00000000\ \mathbf{1}$$

There are times when we can arrange numbers—legitimately—creating a rather unexpected pattern, as is the case when we find the decimal equivalent of the fraction

$$\frac{1}{729} = 0.\overline{0013717421124828532235939643347050754458161865569272}$$
$$\overline{9766803840877914951989026 0631}.$$

The eighty-one-digit decimal equivalent can be arranged in nine groups of nine as follows:

.001 371 742
112 482 853
223 593 964
334 705 075
445 816 186
556 927 297
668 038 408
779 149 519
890 260 631.

Looking up and down the columns formed by this arrangement will reveal quite a few unusual patterns—such as consecutive numbers. The curiosities in arithmetic seem to be unending!

UNUSUAL WAYS TO REPRESENT NUMBERS

Consider the question of how to represent the number 34 by using only 3s. The answer is $33 + \frac{3}{3}$.

Or, represent the number 56 using only 5s, which would be $55 + \frac{5}{5}$. We can represent the number 1,000 by using only 9s as $999 + \frac{9}{9}$. However, for another challenge we might try to represent the numbers from 1 to 100 using only 4s. Don't read further until you try it yourself. Although we provide this here, we would expect a motivated reader to do this for other numbers as well.

$$0 = 44 - 44 = \frac{4}{4} - \frac{4}{4}$$

$$1 = \frac{4+4}{4+4} = \frac{\sqrt{44}}{\sqrt{44}} = \frac{4+4-4}{4}$$

$$2 = \frac{4 \cdot 4}{4+4} = \frac{4-4}{4} + \sqrt{4} = \frac{4}{4} + \frac{4}{4}$$

$$3 = \frac{4+4+4}{4} = \sqrt{4} + \sqrt{4} - \frac{4}{4} = \frac{4 \cdot 4 - 4}{4} = 4 - 4^{4-4}$$

$$4 = \frac{4-4}{4} + 4 = \frac{\sqrt{4 \cdot 4 \cdot 4}}{4} = (4-4) \cdot 4 + 4$$

$$5 = \frac{4 \cdot 4 + 4}{4}$$

$$6 = \frac{4+4}{4} + 4 = \frac{4\sqrt{4}}{4} + 4$$

$$7 = \frac{44}{4} - 4 = \sqrt{4} + 4 + \frac{4}{4} = (4+4) - \frac{4}{4}$$

$$8 = 4 \cdot 4 - 4 - 4 = \frac{4(4+4)}{4} = 4 + 4 + 4 - 4$$

$$9 = \frac{44}{4} - \sqrt{4} = 4\sqrt{4} + \frac{4}{4} = \frac{4}{4} + 4 + 4$$

$$10 = 4 + 4 + 4 - \sqrt{4} = \frac{44 - 4}{4}$$

$$11 = \frac{4}{4} + \frac{4}{.4} = \frac{44}{\sqrt{4} + \sqrt{4}}$$

$$12 = \frac{4 \cdot 4}{\sqrt{4}} + 4 = 4 \cdot 4 - \sqrt{4} - \sqrt{4} = \frac{44 + 4}{4}$$

$$13 = \frac{44}{4} + \sqrt{4}$$

$$14 = 4 \cdot 4 - 4 + \sqrt{4} = 4 + 4 + 4 + \sqrt{4} = \frac{4!}{4 + 4 + 4} = 4! - \left(4 + 4 + \sqrt{4}\right)$$

$$15 = \frac{44}{4} + 4 = \frac{\sqrt{4} + \sqrt{4} + \sqrt{4}}{.4}$$

$$16 = 4 \cdot 4 - 4 + 4 = \frac{4 \cdot 4 \cdot 4}{4} = 4 + 4 + 4 + 4$$

$$17 = 4 \cdot 4 + \frac{4}{4}$$

$$18 = \frac{44}{\sqrt{4}} - 4 = 4 \cdot 4 + 4 - \sqrt{4} = 4 \cdot 4 + \frac{4}{\sqrt{4}} = \frac{4! + 4! + 4!}{4}$$

$$19 = \frac{4 + \sqrt{4}}{.4} + 4 = 4! - 4 - \frac{4}{4}$$

$$20 = 4 \cdot 4 + \sqrt{4} + \sqrt{4} = \left(4 + \frac{4}{4}\right) \cdot 4$$

$$21 = 4! - 4 + \frac{4}{4}$$

$$22 = \frac{4}{4}(4!) - \sqrt{4} = 4! - \frac{((4+4)/4)}{4} = \frac{44}{4} \cdot \sqrt{4} = -4 + \frac{4}{4}$$

$$23 = 4! - \sqrt{4} + \frac{4}{4} = 4! - 4^{4-4}$$

$$24 = 4 \cdot 4 + 4 + 4$$

$$25 = 4! - \sqrt{4} + \frac{4}{4} = 4! + \sqrt{(4+4-4)} = \left(4 + \frac{4}{4}\right)^{\sqrt{4}}$$

$$26 = \frac{4}{4}(4!) + \sqrt{4} = 4! + \sqrt{4+4-4} = 4 + \frac{44}{\sqrt{4}}$$

$$27 = 4! + 4 - \frac{4}{4}$$

$$28 = (4+4) \cdot 4 - 4 = 44 - 4 \cdot 4$$

$$29 = 4! + 4 + \frac{4}{4}$$

$$30 = 4! + 4 + 4 - \sqrt{4}$$

$$31 = \frac{((4 + \sqrt{4})! + 4!)}{4!}$$

$$32 = (4 \cdot 4) + (4 \cdot 4)$$

$$33 = 4! + 4 + \frac{\sqrt{4}}{.4}$$

$$34 = \left(4 \cdot 4 \cdot \sqrt{4}\right) + \sqrt{4} = 4! + \left(\frac{4!}{4}\right) + 4 = \sqrt{(4^4)} \cdot \sqrt{4} + \sqrt{4}$$

$$35 = 4! + \frac{44}{4}$$

$$36 = (4+4) \cdot 4 + 4 = 44 - 4 - 4$$

$$37 = 4! + \frac{\left(4! + \sqrt{4}\right)}{\sqrt{4}}$$

$$38 = 44 - \frac{4!}{4}$$

$$39 = 4! + \frac{4!}{4 \cdot .4}$$

$$40 = (4! - 4) + (4! - 4) = 4 \cdot \left(4 + 4 + \sqrt{4}\right)$$

$$41 = \frac{4! + \sqrt{4}}{.4} - 4!$$

$$42 = 44 - 4 + \sqrt{4} = (4! + 4!) - \frac{4!}{4}$$

$$43 = 44 - \left(\frac{4}{4}\right)$$

$$44 = 44 + 4 - 4$$

$$45 = 44 + \frac{4}{4}$$

$$46 = 44 + 4 - \sqrt{4} = (4! + 4!) - \left(\frac{4}{\sqrt{4}}\right)$$

$$47 = (4! + 4!) - \frac{4}{4}$$

$$48 = (4 \cdot 4 - 4) \cdot 4 = 4 \cdot (4 + 4 + 4)$$

$$49 = (4! + 4!) + \frac{4}{4}$$

$$50 = 44 + \left(\frac{4!}{4}\right) = 44 + 4 + \sqrt{4}$$

$$51 = \frac{(4! - 4 + .4)}{.4}$$

$$52 = 44 + 4 + 4$$

$$53 = 4! + 4! + \frac{\sqrt{4}}{.4}$$

$$54 = 4! + 4! + 4 + \sqrt{4}$$

$$55 = \frac{(4! - 4 + \sqrt{4})}{.4}$$

$$56 = 4! + 4! + 4 + 4 = 4 \cdot (4 \cdot 4 - \sqrt{4})$$

$$57 = \left(\frac{4! - \sqrt{4}}{.4}\right) + \sqrt{4}$$

$$58 = ((4! + 4) \cdot \sqrt{4}) + \sqrt{4} = 4! + 4! + \frac{4}{.4}$$

$$59 = \frac{(4! - \sqrt{4})}{.4} + 4 = \frac{4!}{.4} - \frac{4}{4}$$

$$60 = 4 \cdot 4 \cdot 4 - 4 = \frac{4^4}{4} - 4 = 44 + 4 \cdot 4$$

$$61 = \frac{(4! + \sqrt{4})}{.4} - 4 = \frac{4!}{.4} + \frac{4}{4}$$

$$62 = 4 \cdot 4 \cdot 4 - \sqrt{4}$$

$$63 = \frac{(4^4 - 4)}{4}$$

$$64 = 4\sqrt{4} \cdot 4\sqrt{4} = 4 \cdot (4! - 4 - 4) = (4 + 4) \cdot (4 + 4)$$

$$65 = \frac{(4^4 + 4)}{4}$$

$$66 = 4 \cdot 4 \cdot 4 + \sqrt{4}$$

$$67 = \frac{4! + \sqrt{4}}{.4} + \sqrt{4}$$

$$68 = 4 \cdot 4 \cdot 4 + 4 = \frac{4^4}{4} + 4$$

$$69 = \left(\frac{4! + \sqrt{4}}{.4} \right) + 4$$

$$70 = \frac{(4 + 4)!}{4! \cdot 4!} = 44 + 4! + \sqrt{4}$$

$$71 = \frac{4! + 4.4}{.4}$$

$$72 = 44 + 4! + 4 = 4 \cdot \left(4 \cdot 4 + \sqrt{4} \right)$$

$$73 = \frac{4! \cdot \sqrt{4} + \sqrt{.4}}{\sqrt{.4}}$$

$$74 = 4! + 4! + 4! + \sqrt{4}$$

$$75 = \frac{4! + 4 + \sqrt{4}}{.4}$$

$$76 = (4! + 4! + 4!) + 4$$

$$77 = \left(\frac{4}{.4} \right)^{\sqrt{4}} - 4$$

$$78 = 4 \cdot (4! - 4) - \sqrt{4}$$

$$79 = 4! + \frac{4! - \sqrt{4}}{.4}$$

$$80 = (4 \cdot 4 + 4) \cdot 4$$

$$81 = \left(4 - \left(\frac{4}{4}\right)\right)^4 = \left(\frac{4!}{4\sqrt{4}}\right)^4$$

$$82 = 4 \cdot (4! - 4) + \sqrt{4}$$

$$83 = \frac{4! - .4}{.4} + 4!$$

$$84 = 44 \cdot \sqrt{4} - 4$$

$$85 = \frac{\frac{4! + 4}{.4}}{.4}$$

$$86 = 44 \cdot \sqrt{4} - \sqrt{4}$$

$$87 = 4 \cdot 4! - \frac{4}{.4} = 44\sqrt{4} - i^4$$

$$88 = 4 \cdot 4! - 4 - 4 = 44 + 44$$

$$89 = 4! + \frac{4! + \sqrt{4}}{.4}$$

$$90 = 4 \cdot 4! - 4 - \sqrt{4} = 44 \cdot \sqrt{4} + \sqrt{4}$$

$$91 = 4 \cdot 4! - \frac{-\sqrt{4}}{.4}$$

$$92 = 4 \cdot 4! - \sqrt{4} - \sqrt{4} = 44 \cdot \sqrt{4} + 4$$

$$93 = 4 \cdot 4! - \frac{4}{.4}$$

$$94 = 4 \cdot 4! + \sqrt{4} - 4$$

$$95 = 4 \cdot 4! - \frac{4}{4}$$

$$96 = 4 \cdot 4! + 4 - 4 = 4! + 4! + 4! + 4!$$

$$97 = 4 \cdot 4! + \frac{4}{4}$$

$$98 = 4 \cdot 4! + 4 - \sqrt{4}$$

$$99 = \frac{44}{.44} = 4 \cdot 4! + \frac{\sqrt{4}}{\sqrt{.04}} = \frac{4}{4\%} - \frac{4}{4}$$

$$100 = 4 \cdot 4! + \sqrt{4} + \sqrt{4} = \left(\frac{4}{.4}\right) \cdot \left(\frac{4}{.4}\right) = \frac{44}{.44}$$

A DECEPTIVE NUMBER PROPERTY

An entertaining exercise that one can do with numbers is to try to represent each of our natural numbers as the sum of two or more consecutive natural numbers. For example, here are a few such illustrations:

$$3 = 1 + 2$$
$$5 = 2 + 3$$
$$6 = 1 + 2 + 3$$
$$7 = 3 + 4$$
$$10 = 1 + 2 + 3 + 4$$
$$17 = 8 + 9$$
$$18 = 3 + 4 + 5 + 6 = 5 + 6 + 7$$
$$51 = 6 + 7 + 8 + 9 + 10 + 11$$

Can this be done for all numbers? Here are some more examples:

$$9 = 2 + 3 + 4 = 4 + 5$$
$$11 = 5 + 6$$
$$12 = 3 + 4 + 5$$
$$13 = 6 + 7$$
$$14 = 2 + 3 + 4 + 5$$
$$15 = 4 + 5 + 6 = 7 + 8$$

It appears that some numbers have been left out. If you haven't done so, try to see if you can get powers of 2 to follow this pattern. Might you then be able to draw some conclusion from your findings about when this pattern is not possible?

DE POLIGNAC'S FAULTY CONJECTURE

In 1848 the French mathematician Alphonse de Polignac (1817–1890) conjectured that every odd number greater than 1 can be expressed as the sum of a power of 2 and a prime number. As we look at the list below, we find that this conjecture seems to hold up through the number 125. However, when we try to represent 127 as the sum of a power of 2 and a prime number, we find we cannot do it. Consequently, what seemed like a very nice conjecture did not hold true. It is believed that de Polignac claims to have tried this for the first three million numbers and found only the number 959 as one for which this pattern does not hold true. Yet we see in figure 1.15 that the number 127 does not allow itself to be so expressed. There are sixteen such odd numbers in the first one thousand numbers that do not allow themselves to fit this pattern.

Odd Number	Sum of a Power of 2 and a Prime Number
3	$= 2^0 + 2$
5	$= 2^1 + 3$
7	$= 2^2 + 3$
9	$= 2^2 + 5$
11	$= 2^3 + 3$
13	$= 2^3 + 5$
15	$= 2^3 + 7$
17	$= 2^2 + 13$
19	$= 2^4 + 3$
...	...
51	$= 2^5 + 19$
...	...
125	$= 2^6 + 61$
127	$= ?$
129	$= 2^5 + 97$
131	$= 2^7 + 3$
...	...
241	$= 2^7 + 113$
...	...
999,999	$= 2^{16} + 934,463$

Figure 1.15

Numbers that do not fit de Polignac's conjecture are often referred to as *Polignac numbers*. Mathematicians have searched for Polignac numbers and have looked for patterns. For example, it may be of interest to find 2 consecutive odd numbers that can qualify as Polignac numbers. Here are a few: 905 and 907; 3,341 and 3,343; 3,431 and 3,433. By the way, 905 is the first composite number that is a Polignac number. An ambitious reader may choose to search for other such number pairs. Another neat feature about Polignac numbers is that if you subtract 2 from any of these numbers, you will end up with a composite number, and not a prime.

A CURIOSITY OF SUMS

It is known that for a sequence of consecutive natural numbers, the sum of the cubes is equal to the square of the sum. That is, $1^3 + 2^3 + 3^3 + 4^3 + \cdots + n^3 = \left(1 + 2 + 3 + 4 + \cdots + n\right)^2$. We can demonstrate that with an example:

$1^3 + 2^3 + 3^3 + 4^3 + 5^3 + 6^3 + 7^3 = 1 + 8 + 27 + 64 + 125 + 216 + 343 = 784$

$(1 + 2 + 3 + 4 + 5 + 6 + 7)^2 = 28^2 = 784$

The French mathematician Joseph Liouville (1809–1882) rediscovered a rather neat relationship previously attributed to Nicomachus (ca. 100 CE), namely, that the sum of cubes of a certain set of numbers can also be equal to the square of the sum of these numbers. Furthermore, the sum of the first n cubes is the square of the nth *triangular number*.[24] For example, if we take the sum of the cubes of the first five natural numbers, we can show that this sum is equal to the square of the fifth triangular number, which is also the sum of these five natural numbers:

$1^3 + 2^3 + 3^3 + 4^3 + 5^3 = 1 + 8 + 27 + 64 + 125 = 225 = 15^2 = (1 + 2 + 3 + 4 + 5)^2$.

Remember, 15 is the fifth triangular number.

Another set of numbers can be found in the following way: We select any number and list all the divisors of that number. Then we list the number of divisors that each of these divisors has. That will be our critical set of numbers. For example, suppose we select the number 12. The divisors of 12 are 1, 2, 3, 4, 6, and 12. The number of divisors of each of these divisors (in order) is 1, 2, 2, 3, 4, and 6. We will now apply Liouville's theorem:

$$1^3 + 2^3 + 2^3 + 3^3 + 4^3 + 6^3 = 324$$

$$\left(1 + 2 + 2 + 3 + 4 + 6\right)^2 = 18^2 = 324$$

Amazingly, the two are equal!

SIMPLIFYING CUMBERSOME EXPRESSIONS CLEVERLY

Faced with a rather-cumbersome expression to be evaluated, one typically reaches for a calculator. However, even then the following is a rough task to evaluate.

$$(3 + 1) \cdot (3^2 + 1) \cdot (3^4 + 1) \cdot (3^8 + 1) \cdot (3^{16} + 1) \cdot (3^{32} + 1)$$

With the aid of a calculator—and lots of patience—one would get an approximated answer of $1.717 \cdot 10^{30}$.

A fourteen-year-old student in Berlin (Huyen Nguyen Thi Minh), who was a member of the Society of Gifted Mathematics Students in Berlin, came up with a clever technique to simplify this overwhelming expression. She let the expression be represented by x and got:

$$x = (3 + 1) \cdot (3^2 + 1) \cdot (3^4 + 1) \cdot (3^8 + 1) \cdot (3^{16} + 1) \cdot (3^{32} + 1).$$

Then by multiplying by $(3 - 1)$ [=2], she introduced a term that continuously can be combined with succeeding terms as follows:

$x \cdot (3-1) = (3-1) \cdot (3+1) \cdot (3^2+1) \cdot (3^4+1) \cdot (3^8+1) \cdot (3^{16}+1) \cdot (3^{32}+1)$

$x \cdot (3-1) = (3^2-1) \cdot \qquad\qquad (3^2+1) \cdot (3^4+1) \cdot (3^8+1) \cdot (3^{16}+1) \cdot (3^{32}+1)$

$x \cdot (3-1) = (3^2-1) \cdot (3^2+1) \cdot (3^4+1) \cdot (3^8+1) \cdot (3^{16}+1) \cdot (3^{32}+1)$

$x \cdot (3-1) = (3^4-1) \cdot \qquad\qquad (3^4+1) \cdot (3^8+1) \cdot (3^{16}+1) \cdot (3^{32}+1)$

$x \cdot (3-1) = (3^4-1) \cdot (3^4+1) \cdot (3^8+1) \cdot (3^{16}+1) \cdot (3^{32}+1)$

$x \cdot (3-1) = (3^8-1) \cdot \qquad\qquad (3^8+1) \cdot (3^{16}+1) \cdot (3^{32}+1)$

$x \cdot (3-1) = (3^8-1) \cdot (3^8+1) \cdot (3^{16}+1) \cdot (3^{32}+1)$

$x \cdot (3-1) = (3^{16}-1) \cdot \qquad\qquad (3^{16}+1) \cdot (3^{32}+1)$

$x \cdot (3-1) = (3^{16}-1) \cdot (3^{16}+1) \cdot (3^{32}+1)$

$x \cdot (3-1) = (3^{32}-1) \cdot \qquad\qquad (3^{32}+1)$

$x \cdot (3-1) = (3^{32}-1) \cdot (3^{32}+1)$

$x \cdot (3-1) = 3^{64}-1$

This, then, reads as $2x = 3^{64} - 1$, which allows us to calculate x from:
$$x = \frac{3^{64} - 1}{2}.$$

We, therefore, transformed this cumbersome expression with this clever approach to get

$$(3 + 1) \cdot (3^2 + 1) \cdot (3^4 + 1) \cdot (3^8 + 1) \cdot (3^{16} + 1) \cdot (3^{32} + 1) = \frac{3^{64} - 1}{2}$$
= 1,716,841,910,146,256,242,328,924,544,640, which can also be written as
1.716841910146256242328924544640 · 10^{30}, analogous to the approximation
that the calculator would give us.

ALGEBRA REVEALS A CURIOSITY

We know from algebra that squaring a binomial such as $x+1$ gives us
$$(x+1)^2 = x^2 + 2x + 1 = x^2 + (2x+1)$$
This tells us that to get from one square number, say x^2, to the next
square number, $(x+1)^2$, we simply have to add twice x plus 1 (that is, $2x +$
1), since $(x+1)^2 = x^2 + (2x+1)$. To see how this works, we can take the square
number $4^2 = 16$ and add $2 \cdot 16 + 1 = 33$, then by adding this to the square,
we get the next square number, 16 + 33 = 49 = 7^2.

Thus, the difference between consecutive squares is twice the square
root of the first square plus 1. We can show this with the following example.
The difference between the squares 64 and 81 can then be found by taking
twice the square root of 64 and adding 1—that is, 2 · 8 + 1 = 17, which
is equal to 81 – 64. This gives us a nice procedure of finding the difference between any two square numbers. This is rather easy to see since
$(x+1)^2 = x^2 + 2x + 1 = x^2 + (2x+1)$, which yields the above-mentioned relationship: $(x + 1)^2 - x^2 = 2x + 1$.

PATTERN RECOGNITION OF CONSECUTIVE SQUARES

We began our investigation of consecutive squares by observing the
following pattern:
$$2^2 - 1^2 = \ \ 4 - \ 1 = 3 = 2 + 1$$
$$3^2 - 2^2 = \ \ 9 - \ 4 = 5 = 3 + 2$$
$$4^2 - 3^2 = 16 - \ 9 = 7 = 4 + 3$$
$$5^2 - 4^2 = 25 - 16 = 9 = 5 + 4$$

With a little bit of manipulation we can create from the above equations
the following:

$$2^2 - 2 = 1^2 + 1$$
$$3^2 - 3 = 2^2 + 2$$
$$4^2 - 4 = 3^2 + 3$$
$$5^2 - 5 = 4^2 + 4$$

The question one has to ask oneself is, will this be true for all such squares?

As we ponder this question, let us take a larger number to see what may occur:

$$25^2 - 24^2 = 625 - 576 = 49 = 25 + 24.$$

This can be rewritten as $25^2 - 25 = 24^2 + 24$.

It would *seem* that this pattern does hold for all numbers, but that is by no means a proof that it will be true, in fact, for all numbers. This leads us to considering a general case for the number a, where we would like to determine if $a^2 - (a-1)^2 = \sqrt{a^2} + \sqrt{(a-1)^2} = a + (a-1)$.

A simple proof would go like this:

$a^2 - (a-1)^2 = a^2 - (a^2 - 2a + 1) = a^2 - a^2 + 2a - 1 = 2a - 1 = a + (a-1)$.

We can see, therefore, that the difference of consecutive squares is equal to the sum of the square roots, or put another way, the sum of their bases. At the same time, we also proved that the difference of consecutive squares is always an odd number.[25]

A more ambitious question would be to consider two randomly selected squares and see if there is some structure that can be generalized to the difference of these two squares. In other words, we will look to see if there is a particular pattern resulting from the following equation: $a^2 - b^2 = \sqrt{a^2} + \sqrt{b^2} = a + b$, where a and b are any natural numbers. We begin with $a^2 - b^2 = a + b$.

Then factoring the difference of two squares gives us:

$(a + b)(a - b) = a + b$.

By adding $-(a + b)$ to both sides of the equation, we get:

$(a + b)(a - b) - (a + b) = 0$.

This can be simplified as the following: $(a + b)(a - b - 1) = 0$.

We know that when the product of two numbers is 0, one or both of the two factors must be 0. If both a and b are 0, then clearly this equation is satisfied. However, this trivial case does not interest us much. Once the numbers a and b are greater than zero, then also $a + b$ is greater than zero. Therefore, the above equation is only satisfied if $a - b - 1 = 0$, or $a = b + 1$.

This tells us that the above equation is satisfied whenever a and b differ by 1. What does this tell you about our original conjecture?

AN ALTERNATE WAY OF FINDING PRIME NUMBERS

While the sieve of Eratosthenes (276–ca. 194 BCE) is probably the best-known algorithm for generating primes, in 1934, an Indian student, S. P. Sundaram, developed a sieve of numbers that can also be used to generate prime numbers. This rather-curious technique is not well known, but it works! The sieve (figure 1.16) also has the first row and the first column in an arithmetic progression, where the first term is 4 and the common difference is 3, as follows: 4, 7, 10, 13, 16, 19, ... Each succeeding row beginning with a member of this original arithmetic progression increases its common difference by 2 so that the common difference of row 2 is 5, and the common difference of row 3 is 7, and so on.

4	7	10	13	16	19	22	25	28	...
7	12	17	22	27	32	37	42	47	...
10	17	24	31	38	45	52	59	66	...
13	22	31	40	49	58	67	76	85	...
16	27	38	49	60	71	82	93	104	...
19	32	45	58	71	84	97	110	123	...
...	...	...	...	...	...	...	...	...	

Figure 1.16

Now that we have the table up to a usable point (figure 1.16), we can apply it in the following way: for any number n that is not in the table, the number $2n + 1$ is a prime number. Alternately, for any number n that is in the table, the number $2n + 1$ is *not* a prime number.[26] For example, if we choose the number 33, which is not in the table, then we know that the number $2 \cdot 33 + 1 = 67$ is a prime number. Alternately, if we choose the number 31, which is in the table, the number $2 \cdot 31 + 1 = 63$ is *not* a prime number. Perhaps the small table does not yield many surprises, but just imagine a very large table that would allow you to generate prime numbers with the confidence that they are, in fact, prime numbers.

SURPRISING APPEARANCE OF POWERS

In 1951, Alfred Moessner published a paper[27] with an interesting discovery that we will enjoy here.

We begin with the list of natural numbers and place a box around every second member of the list.

1, [2], 3, [4], 5, [6], 7, [8], 9, [10], 11, [12], 13, [14], 15, [16], 17, [18], . . .

We will then take the sum of the unboxed numbers preceding each boxed number sequentially, as shown below in the second row. In the third row you will then see the sums of the second row, and we have a list of squares appearing there: 1, 4, 9, 16, 25, 36, 49, . . .

1	[2]	3	[4]	5	[6]	7	[8]	9	[10]	11	[12]	13	[14]
1		1+3		1+3+5		1+3+5+7		1+3+5+7+9		1+3+5+7+9+11		1+3+5+7+9+11+13	
1		4		9		16		25		36		49	

15	[16]	17	[18]	19	[20]	...
1+3+5+7+9+11+13+15		1+3+5+7+9+11+13+15+17		1+3+5+7+9+11+13+15+17+19		
64		81		100		

We will now repeat this process, but this time we will take every *third* member of the natural numbers and box them. Then we will take the sums of the unboxed numbers and box the last number in each group (third row), which is the one in the position before the boxed number in the first row. We can continue this another time, and we will find the remaining numbers are the sequence of cubes (boxed).

1	2	[3]	4	5	[6]	7	8	[9]	10	11
1	1+2		1+2+4	1+2+4+5		1+2+4+5+7	1+2+4+5+7+8		1+2+4+5+7+8+10	1+2+4+5+7+8+10+11
1	[3]		7	[12]		19	[27]		37	[48]
1			1+7			1+7+19			1+7+19+37	
[1]			[8]			[27]			[64]	

[12]	13	14	[15]	16	...
	1+2+4+5+7+8+10+11+13	1+2+4+5+7+8+10+11+13+14		1+2+4+5+7+8+10+11+13+14+16	
	61	[75]		91	
	1+7+19+37+61			1+7+19+37+61+91	
	[125]			[216]	

We now continue this process, but this time we will box every *fourth* number. We observe that the end result is the sequence of fourth powers of the natural numbers.

1	2	3	[4]	5	6	7	[8]	9	10	11	[12]	13	14	15	[16]	17	18	19	[20]	...
1	3	[6]		11	17	[24]		33	43	[54]		67	81	[96]		113	131	[150]		...
1	[4]			15	[32]			65	[108]			175	[256]			369	[500]			
[1]				[16]				[81]				[256]				[625]				

Let us now box a different set of members of the natural numbers.

We will box the *triangular numbers*[28]: (0,) 1, 3, 6, 10, 15, 21, . . . We then follow the same procedure as before—totaling the unboxed numbers. The resulting numbers this time are the *factorials*,[29] namely 1!, 2!, 3!, 4!, . . .

[1]	2	[3]	4	5	[6]	7	8	9	[10]	11	12	13	14	[15]	16	17	18	19	20	[21]	...
	[2]		6	[11]		18	26	[35]		46	58	71	[85]		101	118	136	155	[175]		...
			[6]			24	[50]			96	154	[225]			326	444	580	[735]			
						[24]				120	[274]				600	1044	[1624]				
										[120]					720	[1764]					
															[720]						

We shall try this procedure another time, but this time we will box the *square numbers*. This time the resulting numbers are a bit puzzling.

[1]	2	3	[4]	5	6	7	8	[9]	10	11	12	13	14	15	[16]	17	18	19	...
	2	[5]		10	16	23	[31]		41	52	64	77	91	[106]		123	141	160	
	[2]			12	28	[51]			92	144	208	285	[376]			499	640	800	
				12	[40]				132	276	484	[769]				1268	1908	2708	
				[12]					144	420	[904]					2172	4080	6788	
									144	[564]						2736	6816	13604	
									[144]							2880	9696	[23300]	
																2880	12576		
																[2880]			

To make some sense from these resulting numbers, we look back at the previously generated square numbers. They can be represented as

$$1$$
$$1 + 2 + 1$$
$$1 + 2 + 3 + 2 + 1$$
$$1 + 2 + 3 + 4 + 3 + 2 + 1$$
$$1 + 2 + 3 + 4 + 5 + 4 + 3 + 2 + 1.$$

Replacing the addition with multiplication we get

$$1$$
$$1 \cdot 2 \cdot 1$$
$$1 \cdot 2 \cdot 3 \cdot 2 \cdot 1$$
$$1 \cdot 2 \cdot 3 \cdot 4 \cdot 3 \cdot 2 \cdot 1$$
$$1 \cdot 2 \cdot 3 \cdot 4 \cdot 5 \cdot 4 \cdot 3 \cdot 2 \cdot 1.$$

You will notice that the product of the numbers in each of the rows is: 1, 2, 12, 144, 2880, . . . , which are the last boxed numbers above. Continuing along, we have the sequence:

1; 2; 12; 144; 2,880; 86,400; 3,628,800; 203,212,800; 14,631,321,600; 1,316,818,944,000; 144,850,083,840,000; . . .

Readers may wish to find other patterns using this technique.

Analogously, we can create another pattern by considering the alternating sums of factorials:

$n! - (n-1)! + (n-2)! - \ldots 1!$, which provide a pattern to be inspected.[30]

$3! - 2! + 1!$	=	5
$4! - 3! + 2! - 1!$	=	19
$5! - 4! + 3! - 2! + 1!$	=	101
$6! - 5! + 4! - 3! + 2! - 1!$	=	619
$7! - 6! + 5! - 4! + 3! - 2! + 1!$	=	4,421
$8! - 7! + 6! - 5! + 4! - 3! + 2! - 1!$	=	35,899

. . .

These sums are always prime numbers. If we continue this sequence, we get **5, 19, 101, 619, 4421, 35899**, 326981, 3301819, 36614981, . . . We would be quick to conclude that all such sums are prime numbers, but that would be a mistake, since $9! - 8! + 7! - 6! + 5! - 4! + 3! - 2! + 1! = 326,981$ is not a prime number, as we can see with $326,981 = 79 \cdot 4139$.

INTERESTING POWERS OF 2

It is interesting to notice that every positive integer can be written uniquely as the sum of distinct powers of 2.

Here are a few examples:

$$1 = 2^0$$
$$5 = 2^0 + 2^2$$
$$9 = 2^0 + 2^3$$
$$11 = 2^0 + 2^1 + 2^3$$
$$31 = 2^0 + 2^1 + 2^2 + 2^3 + 2^4$$

What we have here is merely an example of the representation of all numbers in the binary system (base 2).[31]

We leave it to the reader to extend this list for other positive integers.

Here we offer some powers of 2, where the powers are themselves powers of 2:

$$2^1 = 2$$
$$2^2 = 4$$
$$2^4 = 16$$
$$2^8 = 256$$
$$2^{16} = 65,536$$
$$2^{32} = 4,294,967,296$$
$$2^{64} = 18,446,744,073,709,551,616$$
$$2^{128} = 340,282,366,920,938,463,463,374,607,431,768,211,456$$

$2^{256} = 115,792,089,237,316,195,423,570,985,008,687,907,853,269,984,$
$665,640,564,039,457,584,007,913,129,639,936$

$2^{512} = 13,407,807,929,942,597,099,574,024,998,205,846,127,479,365,$
$820,592,393,377,723,561,443,721,764,030,073,546,976,801,874,298,169,$
$903,427,690,031,858,186,486,050,853,753,882,811,946,569,946,433,649,$
$006,084,096$

By the way, the value of 2^{86}, when expanded in decimal form, will be a number containing no zeros. It is conjectured that this may be the largest power of 2 that has this characteristic. Readers may wish to use a computer to verify this by raising the number 2 to the 86th power.

A PATTERN OF POWERS OF 2

As we show in figure 1.17, powers of 2 can have various initial digits. For example, when we take the first nine powers of 2 and take note of their initial digits, we have all the digits except for 7 and 9 represented as initial digits. Yet when we take 2^{46} and 2^{53}, we find that these powers of 2 give us these remaining two digits.

Power (Exponent)	Power of 2	Initial Digit
1	2	2
2	4	4
3	8	8
4	16	1
5	32	3
6	64	6
7	128	1
8	256	2
9	512	5
. . .		
46	70,368,744,177,664	7
53	9,007,199,254,740,992	9

Figure 1.17

In other words, there will always be at least one power of 2 that begins with one of the nine digits—as we have shown above. This curious property is not reserved only for single-digit beginnings but rather for any number of digit-combination beginnings. That is, if we select the initial digits of a number to be 262, then we can find at least one power of 2 that will give us these initial three digits—in this case $2^{18} = 262,144$. As if this isn't by itself spectacular enough, we can extend this curious property to powers other than 2—yet *not* to any power of 10. Suppose we choose to find a power of 7 whose initial digits are 1628, we will find that it does exist, namely, $7^{18} = 1,628,413,597,910,449$. The proof of this remarkable relationship is beyond the scope of this book. However, an ambitious reader can find a clever proof in Ross Honsberger's *Ingenuity in Mathematics* (Yale University, 1970), pages 38–45.

THE SUM OF THE SQUARES OF THE DIGITS

As we begin to consider the sum of the squares of number, we must take note of the number 110, as it can be represented as the sum of squares in precisely three ways:

$$110 = 1^2 + 3^2 + 10^2 = 1 + 9 + 100$$
$$110 = 5^2 + 6^2 + 7^2 = 25 + 36 + 49$$
$$110 = 2^2 + 5^2 + 9^2 = 4 + 25 + 81$$

Now, moving along to other numbers, there are curious properties of numbers that takes a bit of time to produce but will leave you in awe. We begin with a randomly selected four-digit number, for example, the number 1,527. We then find the sum of the squares of the digits, in this case it is $1^2 + 5^2 + 2^2 + 7^2 = 1 + 25 + 4 + 49 = 79$. We now continue this process by taking the sum of the squares of the digits of the number 79. That is, $7^2 + 9^2 = 130$. Once again, repeating this process we get the following: $1^2 + 3^2 + 0^2 = 1 + 9 + 0 = 10$. The continued process then yields: $1^2 + 0^2 = 1$. What you will notice is that regardless of which number you select at the beginning, this process of the sum of the squares of the digits of each resulting number will end up with either the number 1, as was the case in our illustration above, or the number 4, in which case the succeeding generated numbers after 4 will be the sequence: 4, 16, 37, 58, 89, 145, 42, 20, **4, 16, 37, 58, 89, 145, 42, 20,** 4, 1,6, 37, 58, 89, 145, 42, 20. You will have noticed that the sequence repeats, beginning with the number 4.

For which natural numbers n will the loop [4, 16, 37, 58, 89, 145, 42, 20 (, 4)] of length 8, or the loop [1] of length 1, will result, is a very interesting question. This can be found as a more-complete treatment for the numbers 1–100, and 9,990–10,000 in *Mathematical Amazements and Surprises.*[32]

There is yet another form of taking the powers of the digits of a number—beyond just considering the square of the digits. This time we will take the digits of a number to the power represented by the number itself. So that for the number 3,435, we find (surprisingly) that $3,435 = 3^3 + 4^4 + 3^3 + 5^5$. Numbers that have this property are often referred to as *Münchhausen numbers*. Recognizing that 0^0 is not well defined, yet if we define $0^0 = 0$ for our purposes here, then there are exactly four Münchhausen numbers: 0; 1; 3,435; and 438,579,088.

Here we have these two expanded as described above:
$$3,435 = 3^3 + 4^4 + 3^3 + 5^5 = 27 + 256 + 27 + 3,125$$
$$438,579,088 = 4^4 + 3^3 + 8^8 + 5^5 + 7^7 + 9^9 + 0^0 + 8^8 + 8^8$$
$$= 256 + 27 + 16,777,216 + 3,125 + 823,543 + 387,420,489 + 0 + 16,777,216 + 16,777,216.$$

There are other numbers that can be expressed as the sum of powers of its digits yet, not as above, where the power and the base are the same—rather where the powers are the first few natural numbers. See below:

$$43 = 4^2 + 3^3$$
$$63 = 6^2 + 3^3$$
$$89 = 8^1 + 9^2$$
$$135 = 1^1 + 3^2 + 5^3$$
$$175 = 1^1 + 7^2 + 5^3$$
$$518 = 5^1 + 1^2 + 8^3$$
$$598 = 5^1 + 9^2 + 8^3$$
$$1306 = 1^1 + 3^2 + 0^3 + 6^4$$
$$1676 = 1^1 + 6^2 + 7^3 + 6^4$$
$$2427 = 2^1 + 4^2 + 2^3 + 7^4$$

LUCKY NUMBERS

Mathematicians often search for unusual patterns and come up with some unusual relationships. One can ask, how does that move our understanding of mathematics ahead? The answer is sometimes enigmatic, in that the findings are rather limited. Such is the case with a series of numbers called *lucky numbers*. In 1956, the Polish-American mathematician Stanislaw Marcin Ulam (1909–1984) and others wrote[33] about the idea of setting up a sieve of numbers similar to what Eratosthenes did to generate prime numbers, but instead they used a different procedure yet also began with the natural numbers. The lore is that the idea came from Flavius Josephus (37/38–ca. 100), who told the story of the Siege of Yodfat, where he and his forty soldiers were trapped in a cave, to which the exit was blocked by the Romans. Rather than be captured, they chose suicide—forming a circle and counting off in intervals of three and continuing along with these 3-intervals until there was only one person left—the lucky person. Through luck, he and another man remained and were captured by the Romans. A theoretical counting-out game is often referred to as the Josephus Problem (or Josephus Permutation).

We will do that now with the natural numbers up to the number 20, deleting every second number.

1, 2, 3, 4, 5, 6, 7, 8, 9, 10, 11, 12, 13, 14, 15, 16, 17, 18, 19, 20

1, 2, 3, 4, 5, 6, 7, 8, 9, 10, 11, 12, 13, 14, 15, 16, 17, 18, 19, 20

That is, all the even numbers were deleted. What remained were the odd numbers: 1, 3, 5, 7, 9, 11, 13, 15, 17, 19.

The next number not touched is the 3, so the plan is to delete every third number, as shown here: 1, 3, 5, 7, 9, 11, 13, 15, 17, 19.

This, then, leaves us with the following sequence: 1, 3, 7, 9, 13, 15, 19.

The next untouched number is the number 7. Therefore, we will now take every seventh number and eliminate it—in this case, the number 19 is removed: 1, 3, 7, 9, 13, 15, 19.

This leaves us with the following sequence: 1, 3, 7, 9, 13, 15.

At this point, we have eliminated all the unlucky numbers up to the number 20. This means that the remaining numbers are called the *lucky numbers*—referring back to those lucky soldiers who were not asked to commit suicide. Had we continued—that is, had we taken numbers beyond 20 to determine more lucky numbers—we would have the following sequence:

$L := \{ l \mid l \text{ is } lucky \} = \{1, 3, 7, 9, 13, 15, 21, 25, 31, 33, 37, 43, 49, 51, 63, 67, 69, 73, 75, 79, 87, 93, 99, \ldots\}$

Although mathematicians have spent time trying to determine the significance of these numbers, a few properties have already turned up to present themselves. For example, there are infinitely many lucky numbers. We also have numbers that we can call *lucky primes*, which are lucky numbers that are also prime numbers. To date, it is not known whether there are infinitely many lucky primes. The first few are 3, 7, 13, 31, 37, 43, 67, 73, 79, 127, 151, 163, and 193.

There are other numbers beyond lucky numbers, which we will now consider.

HAPPY AND UNHAPPY NUMBERS

Here we have another rather-strange curiosity in our decimal number system that leads us to partition our numbers into two categories. We say that a number is considered a *happy number* if we take the sum of the squares of the digits to get a second number, and then take the sum of the squares of the digits of that second number to get a third number, and then take the sum of the squares of the digits of the third number, and then continue until you get to the number 1. The numbers where this scheme is applied and do *not* eventually end with the number 1 are called *unhappy numbers*. These latter numbers will eventually end up in a loop—where they will continue through a cycle.

Let's consider one of the happy numbers, say 13, and follow this process of taking the sum of the squares of the digits continuously.

$$1^2 + 3^2 = 10$$
$$1^2 + 0^2 = 1$$

Let's consider a slightly longer path to the number 1 by using 19.

$$1^2 + 9^2 = 82$$
$$8^2 + 2^2 = 68$$
$$6^2 + 8^2 = 100$$
$$1^2 + 0^2 + 0^2 = 1$$

Following is a list of the happy numbers from 1 to 1,000. You might want to try a few of these to test their "happiness": 1, 7, 10, 13, 19, 23, 28, 31, 32, 44, 49, 68, 70, 79, 82, 86, 91, 94, 97, 100, 103, 109, 129, 130, 133, 139, 167, 176, 188, 190, 192, 193, 203, 208, 219, 226, 230, 236, 239, 262, 263, 280, 291, 293, 301, 302, 310, 313, 319, 320, 326, 329, 331, 338, 356, 362, 365, 367, 368, 376, 379, 383, 386, 391, 392, 397, 404, 409, 440, 446, 464, 469, 478, 487, 490, 496, 536, 556, 563, 565, 566, 608, 617, 622, 623, 632, 635, 637, 638, 644, 649, 653, 655, 656, 665, 671, 673, 680, 683, 694, 700, 709, 716, 736, 739, 748,

761, 763, 784, 790, 793, 802, 806, 818, 820, 833, 836, 847, 860, 863, 874, 881, 888, 899, 901, 904, 907, 910, 912, 913, 921, 923, 931, 932, 937, 940, 946, 964, 970, 973, 989, 998, 1,000.

The first few consecutive happy numbers (n; $n + 1$) are: (31, 32), (129, 130), (192, 193), (262, 263), (301, 302), (319, 320), (367, 368), (391, 392), . . .

As you will see when you try to discover happy numbers, many will fall into similar paths, such as the numbers 19 and 91. Following are the unique combinations, that is, none of these numbers will take the same path as they get to the number 1. Naturally, the above list includes variations of these: 1, 7, 13, 19, 23, 28, 44, 49, 68, 79, 129, 133, 139, 167, 188, 226, 236, 239, 338, 356, 367, 368, 379, 446, 469, 478, 556, 566, 888, 899.

It appears that happy numbers are without zeros and with digits in increasing order.

An example of an unhappy number is 25. Notice how this procedure will lead us into a loop and never to reach the number 1.

$$2^2 + 5^2 = 29$$
$$2^2 + 9^2 = 85$$
$$8^2 + 5^2 = \underline{89}$$
$$8^2 + 9^2 = 145$$
$$1^2 + 4^2 + 5^2 = 42$$
$$4^2 + 2^2 = 20$$
$$2^2 + 0^2 = 4$$
$$4^2 = 16$$
$$1^2 + 6^2 = 37$$
$$3^2 + 7^2 = 58$$
$$5^2 + 8^2 = \underline{89}$$

This then begins another loop—repeating everything again with 89, which was reached earlier in this process.

Among the happy numbers there are those that are prime numbers as well—known as *happy prime numbers*. The happy prime numbers less than 500 are the following: 7, 13, 19, 23, 31, 79, 97, 103, 109, 139, 167, 193, 239, 263, 293, 313, 331, 367, 379, 383, 397, 409, 487.

To date, mathematicians have made some claims as to some special happy numbers. For example, the smallest happy number that exhibits all the digits 0–9 is 10,234,456,789. While the smallest happy number that has no zeros, and yet has all the other digits is 1,234,456,789. Then there is the smallest happy number that has no zeros and is also palindromic: 13,456,789,298,765,431. When we accept the inclusion of zeros, the smallest happy number with all the digits is 1,034,567,892,987,654,301. Thus far, the largest happy number where no digit is repeated is 986,543,210.

Looking back in this chapter, we recall the second and third repunit primes,[34] r_{19} = 1111111111111111111, and r_{23} = 11111111111111111111111, which just happen to be happy primes.

An ambitious reader may find other special characteristics of happy numbers.

We can also find happy numbers that form Pythagorean triples.[35] Those so far found (less than 10,000) are as shown in figure 1.18.

(700, 3465, 3535)	(748, 8211, 8245)	(910, 8256, 8306)	(940, 2109, 2309)
(940, 4653, 4747)	(1092, 1881, 2175)	(1323, 4536, 4725)	(1527, 2036, 2545)
(1785, 3392, 3833)	(1900, 1995, 2755)	(1995, 4788, 5187)	(2715, 3620, 4525)
(2751, 8360, 8801)	(2784, 6440, 7016)	(3132, 7245, 7893)	(3135, 7524, 8151)
(3290, 7896, 8554)	(3367, 3456, 4825)	(3680, 5313, 6463)	(4284, 5313, 6825)
(4633, 5544, 7225)	(5178, 6904, 8630)	(5286, 7048, 8810)	(5445, 6308, 8333)
(5712, 7084, 9100)	(6528, 7480, 9928)		

Figure 1.18

Had we extended our earlier list of happy numbers beyond 1,000, we would have reached the first three consecutive happy numbers. They are 1,980; 1,981; and 1,982. If we take the first of these three consecutive happy numbers, we notice that by subtracting the reversal of the number, we end up with a difference using all the same digits as the original number: $1,980 - 0891 = 1,089$.

This number, 1,089, as we will show, has many unique features. However, before we exhibit these amazing features, we already notice an interesting curiosity of this number in that it can also be generated using the same technique with the number 9,108, as follows: $9,108 - 8,019 = 1,089$.

Before we enjoy the additional unique features of the number 1,089, we should note that there are four other four-digit numbers that share this property, namely that by subtracting the number from its reversal, the resulting difference uses all the same digits as the original number.

$$5,823 - 3,285 = 2,538$$
$$3,870 - 0783 = 3,087$$
$$2,961 - 1,692 = 1,269$$
$$7,641 - 1,467 = 6,174$$

The last of these four subtractions leads us to the number 6,174, which we referred to earlier as the *Kaprekar constant*, as we appreciate, its other unique property that we saw under the section "Some Curiosities about Large Numbers."

We would have to go quite a bit further to find the first list of five consecutive happy numbers. To make matters easy, we provide them here: 44,488; 44,489; 44,490; 44,491; and 44,492.

We will leave other characteristics of happy numbers for the reader to find!

THE MYSTICAL NUMBER 1,089

Coming back now to the number 1,089, we realize that some numbers provided us with a great deal of fascination. One such number is 1,089, which we noted earlier in our two subtraction illustrations. Another characteristic of this number can be seen by taking the reciprocal of this number (1,089) and getting the following: $\frac{1}{1089} = .\overline{0009182736455463728191}$.

With the exception of the first three zeros and the last 1, we have a palindromic number, 918,273,645,546,372,819, since it reads the same in both directions. Furthermore, when we multiply 1,089 by 5, we also get a palindromic number 5,445, and if we multiply 1,089 by 9, we get 9,801—the reverse of the original number.

The only other number of four or fewer digits whose multiple is the reverse of the original number is 2,178, since $2,178 \cdot 4 = 8,712$. By the way, 1,089 is also a perfect square, since $33 \cdot 33 = 1,089 = 11^2 \cdot 3^3$.

Let us now do multiplication by 9 of some numbers that are modifications of 1,089, say 10,989; 109,989; 1,099,989; 10,999,989, and so on, and then marvel at the results.

$$10989 \cdot 9 = 98901$$
$$109989 \cdot 9 = 989901$$
$$1099989 \cdot 9 = 9899901$$
$$10999989 \cdot 9 = 98999901$$

and so on.

Now let's get back to 1,089. There is a neat little trick that we can do with 1,089. Suppose you select any three-digit number whose units digit and hundreds digits are not the same, and then reverse that number. Now subtract the two numbers you have (obviously, the larger minus the smaller). Once again reverse the digits of this difference, and add then this new number to the difference. The result will always be 1,089. To see how this works, we will choose any randomly selected three-digit number, say 732. We now subtract $732 - 237 = 495$. Reversing the digits of 495, we get 594, and now we add these last two numbers: $495 + 594 = 1,089$. Yes, this will hold true for all such three-digit numbers—amazing! This is a cute little "trick" that can be justified with simple algebra.

Here is the justification using elementary algebra. We shall represent the arbitrarily selected three-digit number, $\overline{htu}$, as $100h + 10t + u$, where h represents the hundreds digit, t represents the tens digit, and u represents the units digit. The number with the digits reversed is then $100u + 10t + h$. We will let $h > u$, which would be the number you selected or its reverse. In the subtraction, $u - h < 0$; therefore, take 1 from the tens place (of the minuend) making the units place $10 + u$. Since the tens digits of the two numbers to be subtracted are equal, and 1 was taken from the tens digit of the minuend, then the value of this digit is $10(t - 1)$. The hundreds digit of the minuend is $h - 1$, because 1 was taken away to enable subtraction in the tens place, making the value of the tens digit $10(t - 1) + 100 = 10(t + 9)$.

When we do the first subtraction, we get:

$$
\begin{array}{rlll}
 & 100(h-1) & +\quad 10(t+9) & +\quad (u+10) \\
- & (100u & +\quad 10t & +\quad h\,) \\
\hline
 & 100(h-u-1) & +\quad 10 \cdot 9 & +\quad u-h+10
\end{array}
$$

Reversing the digits of this difference $100(h-u-1)+10\cdot 9+(u-h+10)$ gives us: $100(u-h+10)+10\cdot 9+(h-u-1)$.

By adding these last two expressions, we get:
$100(h-u-1)+10\cdot 9+(u-h+10)+100(u-h+10)+10\cdot 9+(h-u-1)$
$= 1{,}000 + 90 - 1 = \mathbf{1{,}089}$.

This algebraic justification enables us to inspect the general case of this arithmetic process, allowing us to guarantee that this process holds true for all numbers.

SOME STRIKINGLY CURIOUS NUMBER RELATIONSHIPS

Here we have a few astonishing patterns that emanate from the list of natural numbers. In the first one, we will begin by listing the natural numbers and grouping them as follows—note how the number of members of each group increases by one each time:

1,
2, 3,
4, 5, 6,
7, 8, 9, 10,
11, 12, 13, 14, 15,

16, 17, 18, 19, 20, 21,
22, 23, 24, 25, 26, 27, 28, . . .

Beginning with the second group, we will now delete every other group, so that we retain the following:

1,
4, 5, 6,
11, 12, 13, 14, 15,
22, 23, 24, 25, 26, 27, 28, . . .

Taking the sum of these remaining numbers gives us $256 = 4^4$.

Suppose we now take the natural numbers only up to 21. We repeat the process, again, beginning with the second group, we eliminate every other group. This will yield the following: 1,   4, 5, 6,   11, 12, 13, 14, 15. Once again, taking the sum of these remaining numbers, we get: $81 = 3^4$.

From this we can deduce that the sum of the members of the n groups, remaining after we have eliminated every other group, is n^4. The ambitious reader may want to verify this for a larger number of natural numbers.

Grouping the natural numbers in another way can also lead to a rather-astonishing pattern.

We will partition the list of natural numbers into groups of various sizes, and, surprisingly, that turn out to have sums that are powers of 3. In other words, the group sizes will be 1, 3, 9, 27, 81, and so on. (See figure 1.19.)

Number of Members of the Group of Natural Numbers	Groups of Consecutive Natural Numbers	Sum of the Group
1	1	$1 = 3^0$
3	2,3,4	$9 = 3^2$
9	5,6,7,8,9,10,11,12,13	$81 = 3^4$
27	14,15,16,17,18,19,20,21,22,23,24,25,26,27,28, 29,30,31,32,33,34,35,36,37,38,39,40	$729 = 3^6$

Figure 1.19

Notice how the exponents for 3 increased by 2 each time. You might want to try to find out if this also is true (which it will be!) for the next eighty-one members of the natural numbers.

By the way, speaking of powers of 3, we find that 121 is the only square number that is the sum of consecutive powers of 3 beginning with 1. That is, $1 + 3 + 9 + 27 + 81 = 121$. And while we are on the number 121, aside from the number 4, the number 121 is the only square to which you can add 4 and result in a cube ($125 = 5^3$).

Here is a nice little oddity. Consider the four numbers 1, 3, 8, and 120. Based on appearance alone, there is nothing to tie these numbers together in any significant way. However, with some ingenuity, or creativity, we can come up with a rather strange relationship. If we multiply any two of these numbers together and then add 1, we will always end up with a square number. For example, suppose we multiply 3 and 8 to get 24, and then add 1, we get a perfect square, 25. Similarly, if we multiply 8 and 120 and then add 1, we get $961 = 31^2$. It has been shown that there exists no number that can augment this set of four numbers and still preserve this cute little property of generating a square number.

It is often fun to take a number and see what kind of relationship can be constructed based on that number. For example, consider the number 132. At first glance there is nothing obviously significant about this number. However, by trying different relationships and patterns, one might stumble onto the fact that 132 is equal to the sum of all two-digit numbers that can be formed with the digits 1, 3, and 2. That is, $12 + 13 + 21 + 23 + 31 + 32 = 132$. It has been shown that this is the smallest number for which this relationship is true.

Here is another surprising relationship that often goes unnoticed: $12^2 = 144$, and reversing the digits, we get $21^2 = 441$. Another such pair of numbers that bears this relationship is $13^2 = 169$; reversing the digits, we get $31^2 = 961$. You might want to search for other such pairs of numbers.

While considering reversals, here is a simple one that can also be appreciated. $497 + 2 = 499$, and $497 \cdot 2 = 994$. Or you might find this one somewhat entertaining: $12 \cdot 42 = 21 \cdot 24 = 504$. It has been shown that there are thirteen such strange pairs of numbers, the largest of which is $36 \cdot 84$ $= 63 \cdot 48 = 3,024$. You may want to try finding the other eleven such pairs.

Another oddity occurs when we take two two-digit square numbers and place them together to form a four-digit square number. The only nontrivial case where this works is the number 1,681 (placing 4^2 next to 9^2), which is a perfect square: $1,681 = 41^2$. The other obvious cases for which this works are those numbers of the form 1,600; 2,500; 3,600, and so on.

The number of oddities or unusual relationships is boundless. Take, for example, the following addition that leads to a perfect square: $621,770 + 077,126 = 698,896 = 836^2$. This would suffice as something quite unusual, but we can make these numbers even more spectacular by now taking their difference: this difference is $621,770 - 077,126 = 544,644 = 738^2$. Lo and behold, another perfect square is produced!

Here is a relationship that would be very difficult to discover because of the large numbers involved. Consider the number $24,678,050$ $= 2^8 + 4^8 + 6^8 + 7^8 + 8^8 + 0^8 + 5^8 + 0^8$, which is the sum of the eighth-power of the digits of the original number. Another oddity can be seen when we have a number in which the sum of the digits, each taken to the power equal to that digit, then results with the original number. That is the case with the number $438,579,088 = 4^4 + 3^3 + 8^8 + 5^5 + 7^7 + 9^9 + 0^0 + 8^8 + 8^8$, if we (for convenience) once again define $0^0 = 0$.

THE MOST CURIOUS "NUMBER"— OR QUANTITATIVE CONCEPT ∞

Perhaps one of the most curious measures of magnitude is the concept that represents infinity, ∞. When we say that a set has an infinite number of elements, to many people that means it has just a very, very large number of elements. Although that is not incorrect, it is "short-selling" this concept. There are, in fact, orders of magnitude of infinity. For example, the set of all natural numbers {1, 2, 3, 4, 5, . . .} is an infinite set. It is not any larger than the set of all even numbers {2, 4, 6, 8, 10, . . .}. Many find that hard to believe, since the set of natural numbers has all the elements of the set of even numbers as well as all of the odd numbers. That logic would have one believe that the set of natural numbers is twice as large as a set of even numbers. But since the two sets are infinite, this is not true. Counterintuitive as it may seem, we can verify this claim of equality in the following way. To show the equivalence of the number of elements of the two sets, we could argue that for every member of the set of natural numbers there will be an element in the set of even numbers to which it can match, thus making the two sets equal in size. That is, $1 \rightarrow 2$, $2 \rightarrow 4$, $3 \rightarrow 6$, $4 \rightarrow 8$, and so on. Of course, this only works when the sets are *infinitely* large. Clearly, if one takes the set of the first hundred natural numbers, that set is clearly larger than the set of even numbers from 2 to 100. It is this concept of infinity that allows us to make this seemingly counterintuitive claim correctly. One way to create a set larger than the set of natural numbers is to take the set of all subsets of natural numbers, which clearly would be a larger infinite set than the infinite set of natural numbers.

The concept of infinity also causes us discomfort in the geometric realm. Let us consider the following. We begin with a staircase where each of the stairs can be of different sizes—although they could just as well be the same size, and it would not disturb our example. In figure 1.20 we show a staircase where the sum of the vertical portions of each stair is a units, and the sum of the horizontal portions of each stair is b units. In other words, if we wanted to carpet the stairs from point P to point Q, we would require a length of carpeting $a + b$ units long.

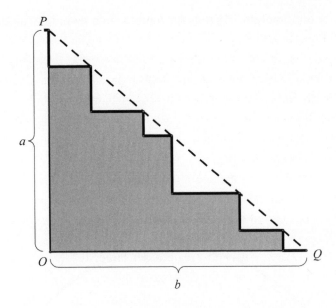

Figure 1.20

In figure 1.20 we can to see that the sum of the bold segments ("stairs"), found by summing all the horizontal and all the vertical segments, is $a+b$. If the number of stairs increases, the sum is still $a+b$, namely the sum of the height OP and the length OQ. Independent of the size of the steps, the sum remains constant: $a + b$. Suppose we keep increasing the number of stairs, and as we do, the stairs naturally become smaller and smaller. At one point, the stairs will become so small that you will not even be able to see them as separate stairs, rather it will look like an oblique plane. Here is where our dilemma arises when we increase the stairs to a "limit" to infinity, so that the set of stairs appears to be a plane—or as seen from a side view, a straight line, which here is the hypotenuse of right triangle POQ. Following this line of reasoning, we would conclude that PQ has length $a + b$. Yet, we know from the Pythagorean theorem that $PQ = \sqrt{a^2 + b^2}$ and *not* $a + b$. So what's wrong?

Nothing is wrong! While the set consisting of the stairs does indeed approach closer and closer to the straight line segment PQ as the number of stairs approaches infinity, it does *not*, however, follow that the *sum* of the bold (horizontal and vertical) lengths approaches the length of PQ, contrary to our intuition. There is no contradiction here, only a failure on the part of our intuition.

One way to explain this dilemma is to argue the following. As the stairs get smaller, they increase in number. In an extreme situation, we have 0-length dimensions (for the stairs) used an infinite number of times, which then leads to considering $0 \cdot \infty$, which is meaningless! Please be warned that this is a very difficult concept, and oftentimes it takes much consideration before acceptance.

We can create a similar argument in another geometric illustration. Consider the semicircles in figure 1.21, where the smaller semicircles extend from one end of the large semicircle's diameter to the other.

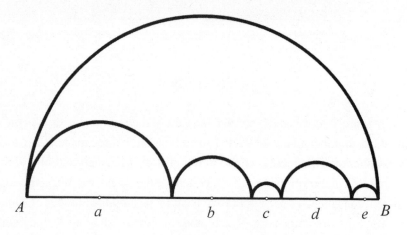

Figure 1.21

It is easy to show that the sum of the arc lengths of the smaller semicircles is equal to the arc length of the larger semicircle. That is, the sum of the smaller semicircles is $\frac{\pi a}{2} + \frac{\pi b}{2} + \frac{\pi c}{2} + \frac{\pi d}{2} + \frac{\pi e}{2} = \frac{\pi}{2}(a+b+c+d+e) = \frac{\pi}{2} \cdot AB$, which is also the arc length of the large semicircle.

As a matter of fact, as we increase the number of smaller semicircles (where, of course, they get smaller) the sum of the semicircle arcs remains the same. No matter how many semicircle arcs we fit in the space between points A and B, the sum of the arcs lengths remains $\frac{\pi}{2} \cdot AB$. As the arcs get smaller and smaller, they will begin to disappear visually, and this arc length sum, $\frac{\pi}{2} \cdot AB$, "appears" to be approaching the length of the segment

AB, but, in fact, it does not! This would be absurd, since we know that $AB \neq \frac{\pi}{2} \cdot AB$, obviously!

Once again, we have a situation that could cause us some unease, since the set of arc lengths—consisting of the smaller semicircles—does indeed appear to approach the length of the straight line segment *AB*. As we have shown above, it does *not* follow, however, that as the *sum* of the semicircles increase indefinitely it approaches the *length* of the limit, in this case *AB*. This is one of the curiosities of the "number" infinity, ∞.

This "apparent limit sum" is absurd, since the shortest distance between points *A* and *B* is the length of segment *AB*, not the semicircle arc *AB* (which equals the sum of the smaller semicircles). This is an important concept to bear in mind so that future misinterpretations involving this curious number, infinity, can be avoided.

HOW RELIABLE IS THE CALCULATOR?

As we move forward in this technologically driven world, our reliance on a calculator becomes more and more unquestioned. Strangely enough, this can lead to some wrong information. Take for example, the following two expressions that need to be evaluated: $\sqrt[12]{1782^{12}+1841^{12}}$ and $\sqrt[12]{3987^{12}+4365^{12}}$.

Taking these one at a time, our calculator returns the following results:

$$\sqrt[12]{1,782^{12}+1,841^{12}} = 1,922$$

$$\sqrt[12]{3,987^{12}+4,365^{12}} = 4,472$$

Yet, for all intents and purposes, we could conclude (based on our calculator results) that $1,782^{12} + 1,841^{12} = 1,922^{12}$ and $3,987^{12} + 4,365^{12} = 4,472^{12}$.

This presents a bit of a dilemma, since we know that the famous French mathematician Pierre de Fermat (1607–1665/66) stated in the margin of one of his algebra books (an edition of Diophantus's *Arithmetica*) in 1637 that the Pythagorean theorem could not be extended beyond the power of 2; in other words, the equation $a^n + b^n = c^n$ does not hold true for natural numbers $n \geq 3$. Although Fermat did not prove his conjecture, stating there was not enough room in the margin of his book page for such a proof, it was eventually proved true by Andrew Wiles (1953–) in 1994. Therefore, it would appear from the above that for the twelfth power it does seem to hold true that $a^n + b^n = c^n$ for $n = 12$. Is the calculator playing a trick on us? The truth be told, when expanded beyond nine decimal places we see that the two numbers are, in fact, different:

$1782^{12} + 1841^{12} =$ **2541210258**6145891762886699581424285266557, but

$1922^{12} =$ **2541210259**3148014108192786496436515657616.

They agree in the first nine places, but that is where the similarity ends. This would leave us a bit skeptical about relying on a calculator to form generalizations. The equality ends with the bold digits, implying that the two are very close in size, but clearly not equal!

A similar argument can be made with the other example:
$3987^{12} + 4365^{12} =$ **63976656349**698612616236230953154487896987106.
However, $4472^{12} =$ **63976656348**486725806862358322168575784124416.
Once again we find that it would be a false conclusion to state that $3987^{12} + 4365^{12} = 4472^{12}$.

In this chapter, we merely attempted to whet the reader's appetite with the boundless curiosities and surprising relationships and patterns that exist in our number system. The only limitation to finding these is our creativity, time, and perhaps the limitations of today's computers.

Having journeyed through a variety of natural numbers, and having discovered boundless curiosities, one might ask, are there any natural numbers that are not interesting? The answer is that all natural numbers are interesting. We can prove this statement: suppose there are some uninteresting natural numbers. Then there must be one that is the smallest uninteresting natural number. This fact already makes this number interesting; therefore, there can be no uninteresting natural numbers. Case closed!

Chapter 2

GEOMETRIC CURIOSITIES

A ROPE AROUND THE EQUATOR

C uriosities in geometry take many forms. They might be visually deceptive, they might lead to unexpected relationships, or they may simply end up counterintuitive. There are times when geometric relationships are truly unanticipated—or even mind-boggling. Take, for example, the situation where a rope is tied along the 40,000-kilometer-long equator of the earth, circumscribing the entire earth's sphere.[1] Now suppose we lengthen this enormously long rope by only 1 meter. It is then no longer tightly tied around the earth. (See figure 2.1.) If we lift this loose rope uniformly around the equator so that it is equally spaced above the equator, would a mouse fit beneath the rope? What does your intuition suggest is the answer?

Figure 2.1   **Figure 2.2**

To analyze this situation, let's focus on figure 2.2, where we picture the two concentric circles—the rope and the circle of the earth. Suppose the sizes of the circles were not given, let's see if we can make some generalization about the distance between the circles. Our question requires us to determine the distance between the two circles ($x = R - r$). First assume (without loss of generality) that the small (inner) circle is extremely small, so small that it has a radius (r) and its circumference (C) of length 0, thus, reducing the inner circle to a point. Then the distance between the circles is merely the radius (R) of the larger circle. We can easily find the circumference of this larger circle using the well-known formula $2\pi R = C + 1$. When we shrink (theoretically, of course) the smaller circle—in this case the earth—to a zero size (i.e., $C = 0$), then the circumference of the larger circle is $2\pi R = 0 + 1 = 1$. The distance between the circles—which is now just the radius of the larger circle—is

$$R = \frac{1}{2\pi} \approx 0.159 \text{meters.}$$

The same result would be found for any size inner circle. Therefore, we can answer the question that 0.159 meters (or about 0.52 feet) is the distance between r and R, and that difference would allow a mouse to comfortably fit beneath the rope. Imagine that by lengthening the rope around the earth by merely 1 meter, we have enough space beneath it to fit a mouse! This should show you that even in geometry not everything is "intuitively obvious" and that there are geometric "facts" that can trick you intuitively as well as optically.

Now that your instinct has been somewhat damaged with the surprising results of the rope around the earth, we present another possible situation. The rope that is 1 meter longer than the circumference of the earth is now to be placed so that it is no longer spaced out over the equator. Rather, it is pulled taut from one external point. (See figure 2.3.) Remember when the rope was equally spaced above the equator, there was merely a 16-centimeter space. Again, what does your intuition suggest is the distance of this external point from Earth? The 1-meter-longer rope pulled taut from a point, where the rest of the rope "hugs" the earth's surface, reaches a point about 122 meters above the earth's surface.

Let's see why this is so. This time the answer is clearly dependent on the size of the earth, and not exclusively on π—but remember π will still play a role in the situation.

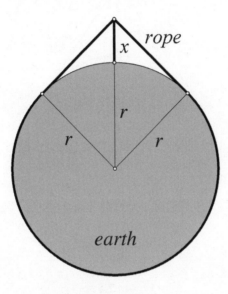

Figure 2.3

This result (that the distance x is approximately 122 millimeters) is perhaps astonishing because one intuitively assumes that if the circumference of the earth is 40,000 kilometers, an extra meter must almost disappear. But this is a mistake! The larger the sphere, the farther the rope can be pulled away from it. For the solution, we need trigonometry and a calculator or a computer.[2]

Referring to figure 2.4, we see that from the exterior point T, the rope (1 meter longer than the circumference of the earth) is pulled taut so that it hugs the earth's surface until the points of tangency (S and Q). We seek to find how high off the surface T is. That means we will try to find the length of x or RT.

Remember the length of the rope from B through S to T is 0.5 meters longer than half the circumference of the earth. So that $\overarc{BS} + \overline{ST} = \overarc{BSR} + \overline{RT} = \overarc{BSR} + 0.5$ meters. Our goal is to find the length of RT (or x).

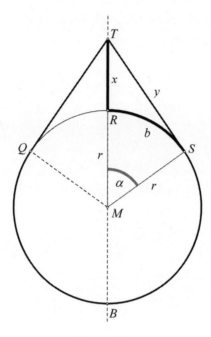

Figure 2.4

So let's review where we are: The rope lies on the arc SBQ, which ends in the points S and Q and at points S and Q goes tangential to the point T. The lengths in the figure above are marked and $\alpha = \angle RMS = \angle RMQ$. (See figure 2.4.)

The length of the rope $+ 1 = 2\pi r + 1$, and we get the following relations: $y = b + 0.5$. This is equivalent to $b = y - 0.5$ (y is 0.5 meters longer than b because of the extension by 1 meter).

In triangle MST (with right angle TSM), the tangent function will be applied: $\tan \alpha = \frac{y}{r}$, therefore, $y = r \cdot \tan \alpha$.

We can form the ratio of arc length to central angle measure and get the following:

$$\frac{b}{\alpha} = \frac{2\pi r}{360°},$$

and then we can get $b = \frac{2\pi \cdot r \cdot \alpha}{360°}$. With $C = 2\pi\, r$, we can find the earth's radius (assuming that the equator is exactly 40,000,000 meters):

$$r = \frac{C}{2\pi} = \frac{40,000,000}{2\pi} \approx 6,366,198 \text{ meters.}$$

Combining the equations we have above, we get the following:

$$b = \frac{2\pi \cdot r \cdot \alpha}{360°} = y - 0.5 = r \cdot \tan \alpha - 0.5.$$

We are now faced with a dilemma. Namely, the equation $\frac{2\pi \cdot r \cdot \alpha}{360°} = r \cdot \tan \alpha - 0.5$ cannot be uniquely solved in the traditional manner because of the number of variables. We will set up a table of possible trial values (figure 2.5) to see what will best satisfy the equation.

$$\frac{2\pi \cdot r \cdot \alpha}{360°} = r \cdot \tan \alpha - 0.5.$$

We will use the value of r we found above: $r = 6,366,198$ meters.

α	$b = \dfrac{2\pi \cdot r \cdot \alpha}{360°}$	$b = r \cdot \tan \alpha - 0.5$	Comparison of Values (Number of Places in Agreement)
30°	3,333,333.478	3,675,525.629	1
10°	1,111,111.159	1,122,531.971	2
5°	555,555.5796	556,969.6547	2
1°	111,111.1159	111,121.8994	4
0.3°	33,333.33478	33,333.13940	5
0.4°	44,444.44637	44,444.66844	5
0.35°	38,888.89057	38,888.87430	6
0.355°	39,444.44615	39,444.45091	6
More exactly:			
0.353°	39,222.22392	39,222.22019	7
0.354°	39 333,33504	39,333.33554	8
0.3545°	39,388.89059	39,388.89322	7
0.355°	39,444.44615	39,444.45091	6

Figure 2.5

Our various trials would indicate that our closest match of the two values occurs at $\alpha \approx 0.354°$. For this value of α, the value of $y = r \cdot \tan \alpha$ $\approx 6,366,198 \cdot 0.006178544171 \approx 39,333.83554$ meters, or about 39,334 meters. The rope is therefore almost 40 kilometers long before it reaches its peak. But how high off the earth's surface is the rope? Or, put another way, what is the length of x?

Applying the Pythagorean theorem to triangle MST we get $MT^2 = r^2 + y^2$. $MT^2 = 6,366,198^2 + 39,333^2 = 40,528,476,975,204 + 1,547,163,556 = 40,530,024,138,760$. Therefore $MT \approx 6,366,319.512$ meters. We are looking for x, which is $MT - r \approx 121.5120192$, or about 122 meters.

This result is perhaps astonishing because one instinctively assumes that, relative to the circumference of the earth (40,000 kilometers), an extra meter must almost disappear. But this is the mistake! Again, the larger the sphere, the farther the rope can be pulled away from it. This surprising result is one reason why mathematics best explains our world!

Looking at the extreme case, where the radius of the equator decreases to zero, we have the minimum value for x, namely, $x = 0.5$ meters.

JAPANESE GEOMETRY—*SANGAKUS*

For over two hundred years, Japan was isolated from the rest of the world in regard to its ideas and culture. During this time, referred to as the Edo period (1603–1868), Euclidean geometry seemed to have blossomed there. Prior to this time, Japanese mathematics was influenced by the Chinese mathematicians. Beginning in about 1603, Western scientific thinking began to emerge in some part because of Japan's commercial interaction with the Netherlands and with Portugal.

By 1631, when Japanese isolation intensified, Japanese mathematics began to become popular as a form of recreation and challenge for the people. They seemed to focus on plane geometry, using triangles, polygons, circles, and circular arcs. In comparison to the European treatment of geometry, which was largely an axiomatic development of the subject, the Japanese were more concerned with the measurements of individual parts of the geometric configuration. A *Sangaku* (a mathematical tablet) was one such medium for spreading these geometric challenges to the people. These tablets (figure 2.6) were artistically presented and were displayed in Shinto shrines and in Buddhist temples. Unsolved problems were presented on these tablets, challenging viewers to seek solutions.

Figure 2.6. *Sangaku* tablet (Ichinoseki Hachiman Shrine No. 1, 1838).

These tablets simply offered problems dealing with applications in basic geometry and no new theorems. However, by solving some of these problems, some new ideas evolved, although much of this was not documented until late in the eighteenth century. In 1790 the Japanese mathematician Fujita Kagen (1765–1821) published the first collection of these *Sangaku* problems in his book *Shimpeki Sampo*, followed by a sequel in 1806 titled *Zoku Shimpeki Sampo*.

The complexity of the problems presented on the *Sangaku* tablets varied from rather simple to more complicated. However, a student with knowledge of today's high-school geometry should be able to solve most of these problems.

In the course of time, the problems presented became more complex. In the eighteenth century, solid geometry problems were included, such as packing spheres inside cones. Through the destruction of some of the Japanese temples, many of these tablets were lost. Today there exist about nine hundred of these *Sangaku* tablets.

The *Sangaku*-tablets problems were accompanied by multicolored figures giving problem details and stating what specific answer was sought. This was followed by the actual answer. However, no method of solution was provided. They were written in *Kanbun*, which was a classical written language with Chinese characters.

Most of the *Sangaku* drawings are very sparse in the information given—as we show in figure 2.7. However, we will provide auxiliary lines and label points as needed (figures 2.8 and 2.9), so that our solutions can be easily followed.

SANGAKU PROBLEM 1

Here we are given a square with the circle inside the square, placed in such a way that is tangent to two sides of the square and to the diagonal of the square (figure 2.7). We are asked to find the radius of the circle in terms of the side of the square. In other words, if we let the radius of the circle have length r, and the side of the square has length a, we would like to get an expression of the value of r in terms of a.

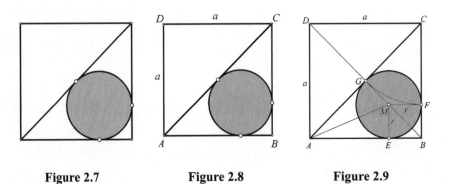

Figure 2.7   Figure 2.8   Figure 2.9

SOLUTION FOR *SANGAKU* PROBLEM 1

The center M of the circle must lie on the diagonal BD, where $r = ME = MF$, and $BG = BM + GM = BM + r$. We know that $MEBF$ is a square with four right angles and two adjacent sides equal. Applying the Pythagorean theorem to triangle BEM, we get

$$BM = \sqrt{ME^2 + MF^2} = \sqrt{2r^2} = r\sqrt{2},$$

$$\text{so that } BG = BM + r = r\sqrt{2} + r = r\left(\sqrt{2} + 1\right).$$

Applying the Pythagorean theorem to triangle BCD, we get

$$BD = \sqrt{BC^2 + CD^2} = \sqrt{2a^2} = a\sqrt{2}.$$

Since the diagonals of the square bisect each other, we have $BG = DG = \frac{BD}{2}$. Combining equalities gives us the following:

$$r\left(\sqrt{2} + 1\right) = \frac{a\sqrt{2}}{2}.$$

Therefore, for the radii $ME = MF = MG = r$, and by rationalizing the denominator, we can conclude that

$$r = \frac{\frac{a\sqrt{2}}{2}}{\sqrt{2} + 1} = \frac{2 - \sqrt{2}}{2} \, a \approx 0.29a.$$

Naturally, there other ways to solve this problem, we merely offered one such here.

SANGAKU PROBLEM 2

We are given a right triangle with an altitude drawn to the hypotenuse, and a circle is inscribed on each side of the altitude, so that each of these circles is tangent to the hypotenuse, to the circumcircle of the right triangle, and to the altitude of the right triangle. (See figures 2.10 and 2.11.) We are being asked to find the radius of each of the two circles in terms of the sides of the right triangle.

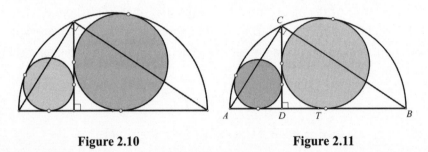

Figure 2.10   Figure 2.11

SOLUTION FOR *SANGAKU* PROBLEM 2

As shown in figure 2.12, point T is the midpoint of diameter AB of the circular arc about triangle ABC, whose sides have lengths a, b, and c.

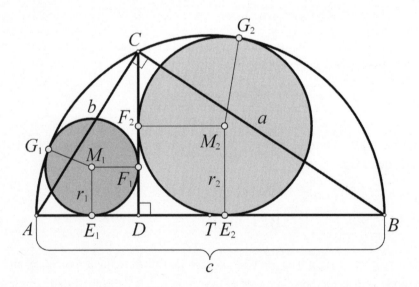

Figure 2.12

It will be sufficient for us to find the radius of the circle shown in figure 2.13, since the circle on the other side of the altitude would be analogous to this one. We will use segment lengths as shown in figure 2.13, where in triangle *ABC*, we have $BD = p$ and $AD = q$. Point *G* is the tangency point of the small circle with the large circular arc. A perpendicular line to the tangent at the point of tangency will travel through the center of the circle. This implies that the points *G*, *M*, and *T* lie on the same line.

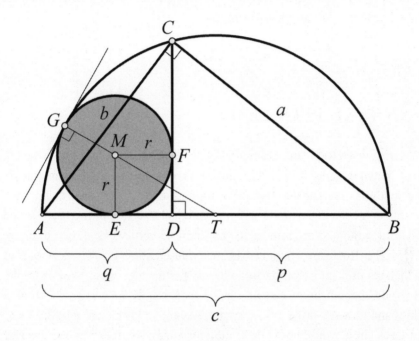

Figure 2.13

We are looking for the length of $EM = r$. We see in figure 2.13 that $ET = ED + DT = ED + (BD - BT) = r + p - \frac{c}{2}$, and $MT = GT - GM = \frac{c}{2} - r$. Now applying the Pythagorean theorem to triangle *EMT* we have $EM^2 + ET^2 = MT^2$, or, using the lengths of these sides, we get the following: $r^2 + (r + p - \frac{c}{2})^2 = (\frac{c}{2} - r)^2$, which then can be written as $r^2 + r^2 + pr - \frac{cr}{2} + pr + p^2 - \frac{cp}{2} - \frac{cr}{2} - \frac{cp}{2} + \frac{c^2}{4} = \frac{c^2}{4} - cr + r^2$, which is simplified as follows:

$$r^2 + 2pr + p^2 = cp, \text{ or } (r + p)^2 = cp.$$

When the altitude is drawn to the hypotenuse of a right triangle, either leg of the right triangle is the mean proportional between the whole hypotenuse and the nearer segment. Therefore, $\frac{AB}{BC} = \frac{BC}{BD}$, or $BC^2 = AB \cdot BD$ and $a^2 = c \cdot p$. By substitution: $(r + p)^2 = a^2$. Since we know the lengths are positive, we get $r + p = a$, or $r = a - p$.

With $p = \frac{a^2}{c}$, it follows that: $r = a - p = a - \frac{a^2}{c} = a\left(1 - \frac{a}{c}\right) = \frac{a(c-a)}{c}$, which is what we sought, namely, to get the value of r in terms of the sides of the triangle. Analogously, we can find the radius of the other circle. Thus, we now have the two radii that we sought:

$$r_1 = a - p = a - \frac{a^2}{c} = a\left(1 - \frac{a}{c}\right) = \frac{a(c-a)}{c}, \text{ and } r_2 = b - q = b - \frac{b^2}{c} = b\left(1 - \frac{b}{c}\right) = \frac{b(c-b)}{c}.$$

SANGAKU PROBLEM 3

We are given a Reuleaux triangle[3] with three congruent circles inscribed in it as shown in figure 2.14. The Reuleaux triangle is very simply constructed by first constructing an equilateral triangle (*ABC*), and then using each vertex as a center of a circle with radius equal to a side of the triangle and drawing a circular arc joining the two remote vertices as shown in figure 2.15. The Reuleaux triangle has many fascinating properties, some of which result from its similarity to the properties of a circle.[4] For example, this odd-shaped figure can be placed tangentially between two parallel lines and turned, all the while remaining tangent to the two parallel lines, just as a circle would under these circumstances. As an example, a circular button fits through a buttonhole regardless of which side of the button you press through buttonhole. The same is true for a button shaped as a Reuleaux triangle, which also fits through a buttonhole regardless which side of the button is pushed through.

The problem here is to determine the radius of these three congruent circles in terms of the side length (*a*) of the equilateral triangle *ABC*.

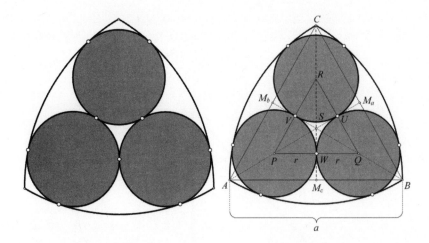

Figure 2.14   **Figure 2.15**

SOLUTION FOR *SANGAKU* PROBLEM 3

In figure 2.15 we are given that the sides of equilateral triangle *ABC* are $AB = BC = AC = a$, and their midpoints are M_c, M_a, and M_b, and their points of tangency are V, U, and W. The radii of the circles are $PW = QW = QU = RU = RV = PV = r$. The common center of both triangles is point S.

By applying the Pythagorean theorem to triangle AM_cC, we get

$$CM_c^2 = AC^2 - AM_c^2 = a^2 - \left(\frac{a}{2}\right)^2 = \frac{3a^2}{4}; \text{ therefore, } CM_c = \frac{a\sqrt{3}}{2}.$$

Similarly we can apply the Pythagorean theorem to triangle *PWR* so that

$$RW^2 = PR^2 - PW^2 = (2r)^2 - r^2 = 3r^2, \text{ and then } RW = r\sqrt{3}.$$

In figure 2.16, we notice that triangle AM_cR is a right triangle with one leg $AM_c = \frac{a}{2}$ and the hypotenuse $AR = AA'' - RA'' = a - r$.

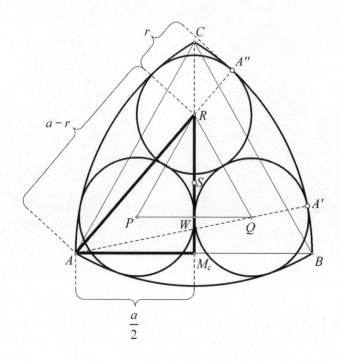

Figure 2.16

Since the medians of a triangle trisect each other, in triangle ABC we have $M_cS = \frac{1}{3}CM_c$, and in triangle PQR we have $SR = \frac{2}{3}RW$. Then applying the results we obtained from the Pythagorean theorem, we get the following:
$M_cR = M_cS + SR = \frac{1}{3} CM_c + \frac{2}{3} RW = \frac{a\sqrt{3}}{6} + \frac{2r\sqrt{3}}{3}$.

When we apply the Pythagorean theorem to triangle AM_cR, we get:

$AR^2 = AM_c^2 + M_cR^2$, also $(a-r)^2 = \left(\frac{a}{2}\right)^2 + \left(\frac{a\sqrt{3}}{6} + \frac{2r\sqrt{3}}{3}\right)^2$, which in turn gives us:

$$a^2 - 2ar + r^2 = \frac{a^2}{4} + \frac{a^2}{12} + \frac{2ar}{3} + \frac{4r^2}{3}$$

$$a^2 - 2ar + r^2 = \frac{1}{3}a^2 + \frac{2}{3}ar + \frac{4}{3}r^2$$

$$\frac{1}{3} \cdot (2a^2 - 8ar - r^2) = 0$$

$$r^2 + 8ar - 2a^2 = 0.$$

Now solving this quadratic equation for r, we get:

$$r_{1,2} = -4a \pm \sqrt{16a^2 + 2a^2} = -4a \pm 3a\sqrt{2},$$

where we ignore the negative root $-4a - 3a\sqrt{2}$.

Therefore the value of the radius length of each of the three circles in terms of the side of equilateral triangle ABC is

$$r = 3a\sqrt{2} - 4a = (3\sqrt{2} - 4) \cdot a \approx 0.24 \cdot a.$$

THE AREAS IN BETWEEN

We are typically accustomed to being asked to find the area of a figure that we can identify by name—such as a circle or a triangle—or a part of a figure. Here we make a departure from this tradition in the *Sangaku* style and seek to find the area between figures, such as between circles. The first of these for consideration is the area between three congruent and mutually tangent circles as shown in figure 2.17.

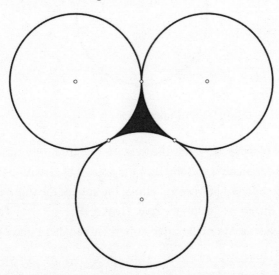

Figure 2.17

In figure 2.18, we joined the centers of the equal circles, whose radius is r, to form the equilateral triangle ABC, whose side has length $2r$. The circles A and B have D as the point of tangency, and A, D, and B are co-linear. We can find the altitude $(CD = h)$ of this triangle using the Pythagorean theorem applied to right triangle ADC: $h^2 + r^2 = (2r)^2$, so that $h = r\sqrt{3}$. We find the area of triangle ABC as follows:

$$A_{\triangle ABC} = 2r \cdot \frac{h}{2} = r \cdot h = r^2\sqrt{3}.$$

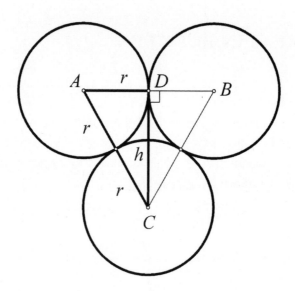

Figure 2.18

To find the space between these three congruent circles, we simply subtract the three shaded sectors from the area of the triangle (figure 2.19). Since each sector has a 60° central angle, the sum of the three sectors is the equivalent of the area of a half a circle. Hence, $A_{3\text{Sectors}} = \frac{\pi r^2}{2}$. Therefore, the area of this space between the three congruent tangent circles is:

$$A_1 = A_{\triangle ABC} - A_{3\text{Sectors}} = r^2\sqrt{3} - \frac{\pi r^2}{2} = r^2\left(\sqrt{3} - \frac{\pi}{2}\right) \approx 0.16\, r^2.$$

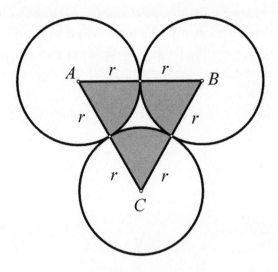

Figure 2.19

Having now found the area encased by the three congruent tangent circles, we will now complicate the situation a bit by inserting a fourth circle within the region between the circles and tangent to each of the larger circles, as shown in figure 2.20, and we want to find the area between this fourth circle and the other three.

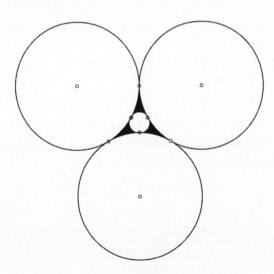

Figure 2.20

We are now faced with two challenges, first, we need to find the radius of the small circle, and then the area encased between the four tangent circles. To do that, we will refer to figure 2.21, which is an enlargement of that shown in figure 2.20, with vertex C missing.

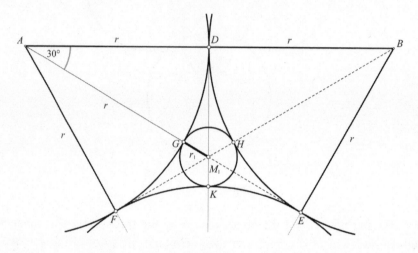

Figure 2.21

Referring to figures 2.18 and 2.21, we notice that the three circles, with centers A, B, and C have tangent points D, E, and F. Since the altitudes, medians, and angle bisectors of an equilateral triangle are all the same lines, and are concurrent at point M_1, we also have (from their median property[5]) that $CM_1 = 2DM_1$, $AM_1 = 2EM_1$, and $BM_1 = 2FM_1$. We will designate the following: $GM_1 = HM_1 = KM_1 = r_1$.

We also have $M_1A = (M_1B = M_1C =) M_1G + GA = r + r_1$.

Furthermore, we know that the three triangles, $\triangle AM_1B$, $\triangle BM_1C$, and $\triangle CM_1A$, are congruent and isosceles, having a vertex angle of $120°$. Thus, for $\triangle ADM_1$, we get the following:

$$\cos \angle BAM_1 = \cos \angle DAM_1 = \tfrac{AD}{AM_1} = \cos 30° = \tfrac{1}{2}\sqrt{3} = \tfrac{r}{r+r_1}.$$

Therefore, $r_1 = \left(\tfrac{2}{\sqrt{3}}-1\right)r = \left(\tfrac{2\sqrt{3}}{3}-1\right)r = \tfrac{2\sqrt{3}-3}{3}\cdot r \ (\approx 0.15 \cdot r)$.

If, on the other hand, you choose to avoid trigonometry, the Pythagorean theorem can be used by applying it to triangle ADM_1, giving us

$$AM_1^2 = AD^2 + DM_1^2, \text{ or } (r + r_1)^2 = r^2 + r^2\left(\frac{\sqrt{3}}{3}\right)^2 = r^2 + \frac{1}{3}\,r^2 = \frac{4}{3}\,r^2.$$

By taking the square root of both sides of this equation, we get $r + r_1 = \frac{2}{\sqrt{3}}\,r$, then, solving for the required radius, $GM_1 = r_1 = \frac{2\sqrt{3}-3}{3}\,r$ $(\approx 0.15 \cdot r)$.

In order to get the required area, A_2, between the four congruent circles, we need to subtract the area of the smallest circle from the area (previously found) between the three larger circles.

The area of smallest circle, $A_{\text{circle}} = \pi \cdot r_1^2 = \pi \cdot \left(\frac{2\sqrt{3}-3}{3} \cdot r\right)^2 = \frac{7-4\sqrt{3}}{3} \cdot \pi \cdot r^2 \approx 0.075 \cdot r^2.$

This enables us to find the area between the four circles by subtracting this smallest circle's area from the area between the three larger circles:

$$A_2 = A_1 - A_{\text{circle}} = \left(\sqrt{3} - \frac{\pi}{2}\right) \cdot r^2 - \frac{7-4\sqrt{3}}{3} \cdot \pi \cdot r^2 = \left(\frac{8\sqrt{3}-17}{6} \cdot \pi + \sqrt{3}\right) r^2 \approx 0.086 \cdot r^2.$$

To complicate matters a bit further, as we inspect the area between circles, we have inserted three even-smaller circles, each of which is tangent to two of the larger circles and the previous small circle as shown in figure 2.22. We designate the radius of each of these smallest circles as r_2.

Our objective here is to find the area between these four circles in terms of the radius of the larger circles, r.

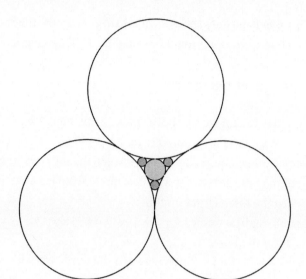

Figure 2.22

We show this in figure 2.23 (an enlargement of a section of figure 2.22) along with several auxiliary lines to help us find the sought-after area.

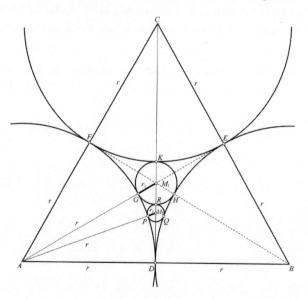

Figure 2.23

To find the length of the radius, r_2, of each of the three smallest circles, we will refer back to our most recent finding, namely that $r_1 = \frac{2\sqrt{3}-3}{3} \cdot r$ ($\approx 0.15 \cdot r$). We also note that the points of tangency of this smallest circle, with center at M_2 and radius r_2, are P, Q, and R. We have then $PM_2 = QM_2 = RM_2 = r_2$ and $RM_1 = r_1$.

To get the radius of the smaller circle, we will consider the two right triangles ADM_1 and ADM_2, where $AM_1 = r + r_1$, and $AM_2 = r + r_2$, as well as $M_1M_2 = M_1R + RM_2 = r_1 + r_2$, so that we can apply the Pythagorean theorem as follows:

$$DM_1 = \sqrt{AM_1^2 - AD^2} = \sqrt{(r+r_1)^2 - r^2} = \sqrt{2rr_1 + r_1^2} = \sqrt{r_1(2r+r_1)}$$

$$= \sqrt{r\left(\frac{2\sqrt{3}}{3}-1\right)\left[2r+r\left(\frac{2\sqrt{3}}{3}-1\right)\right]} = \sqrt{r\left(\frac{2\sqrt{3}}{3}-1\right)\cdot r\left(\frac{2\sqrt{3}}{3}+1\right)} = r\sqrt{\frac{4\cdot 3}{9}-1} = \frac{\sqrt{3}}{3}\,r \qquad (1)$$

($\approx 0.58 \cdot r$).

We could actually get the same result in perhaps a slightly quicker way by considering the centroid M_1 (the point of intersection of the medians) of triangle ABC.

Since $CM_1 = 2DM_1$, we have $DM_1 = \frac{1}{3}\,CD$; however, since CD is the altitude of the triangle, its length is $r\sqrt{3}$; therefore, $DM_1 = \frac{\sqrt{3}}{3}\,r$.

Analogous to our calculation for the length of DM_1, we get the following:

$$DM_2 = \sqrt{AM_2^2 - AD^2} = \sqrt{(r+r_2)^2 - r^2} = \sqrt{2rr_2 + r_2^2} = \sqrt{r_2(2r+r_2)}.$$

Then $DM_1 = DM_2 + M_2M_1 = \sqrt{r_2(2r+r_2)} + r_1 + r_2$

$$= \sqrt{r_2(2r+r_2)} + \left(\frac{2}{\sqrt{3}}-1\right)r + r_2. \qquad (2)$$

Combining the results (1) and (2), we then get the equation:

$$\frac{\sqrt{3}}{3}\,r = \sqrt{r_2(2r+r_2)} + \left(\frac{2}{\sqrt{3}}-1\right)r + r_2.$$

Rearranging the equation gives us:

$$\frac{\sqrt{3}}{3}\,r - \left(\frac{2}{\sqrt{3}}-1\right)r - r_2 = \sqrt{r_2(2r+r_2)}$$

$$\left(\frac{\sqrt{3}}{3}-\frac{2}{\sqrt{3}}+1\right)r - r_2 = \sqrt{r_2(2r+r_2)}$$

$$\left(\frac{\sqrt{3}-2\sqrt{3}+3}{3}\right)r - r_2 = \sqrt{r_2(2r+r_2)}$$

$$\frac{3-\sqrt{3}}{3}\cdot r - r_2 = \sqrt{r_2(2r+r_2)}.$$

Squaring both sides of this equation we get:

$$\frac{(3-\sqrt{3})^2}{9}\cdot r^2 - \frac{2(3-\sqrt{3})}{3}\cdot rr_2 + r_2^2 = 2rr_2 + r_2^2.$$

Adding to both sides of the equation the following: $-r_2^2 + \frac{2(3-\sqrt{3})}{3}\,rr_2$

leaves us with: $\frac{(3-\sqrt{3})^2}{9}\cdot r^2 = 2rr_2 + \frac{2(3-\sqrt{3})}{3}\cdot rr_2.$

Now multiplying both sides of the equation by 9, we get:

$$(3-\sqrt{3})^2r^2 = 18rr_2 + 6(3-\sqrt{3})rr_2.$$

Dividing both sides by r leaves us with:

$$(3-\sqrt{3})^2r = 18r_2 + 6(3-\sqrt{3})r_2$$

$$(9-6\sqrt{3}+3)r = 18r_2 + 18r_2 - 6\sqrt{3}\,r_2$$

$$(12-6\sqrt{3})r = (36-6\sqrt{3})r_2$$

$$r_2 = \frac{12-6\sqrt{3}}{36-6\sqrt{3}} \cdot r = \frac{6(2-\sqrt{3})}{6(6-\sqrt{3})} \cdot r = \frac{2-\sqrt{3}}{6-\sqrt{3}} \cdot r = \left(\frac{3}{11} - \frac{4}{33}\sqrt{3}\right) \cdot r = \frac{9-4\sqrt{3}}{33} \cdot r \ (\approx 0.063 \cdot r).$$

As another way of grappling with this problem, we could also reach back to the well-known law of cosines and apply it to triangle AM_1M_2, so that we can finally get radius r_2:

$$AM_1^2 = AM_2^2 + M_1M_2^2 - 2 \cdot AM_2 \cdot M_1M_2 \cdot \cos \angle AM_2M_1 \text{, so that}$$
$$(r+r_2)^2 = (r+r_1)^2 + (r_1+r_2)^2 - 2 \cdot (r+r_1) \cdot (r_1+r_2) \cdot \cos 60°,$$

(we know that $\cos 60° = \frac{1}{2}$).

We now have: $(r+r_2)^2 = (r+r_1)^2 + (r_1+r_2)^2 - (r+r_1) \cdot (r_1+r_2)$.
Substituting for $r_1 = \frac{2\sqrt{3}-3}{3} \cdot r$, we then get
$$(r+r_2)^2 = -\frac{1}{3} \cdot [(2\sqrt{3}-7)r^2 + 2\sqrt{3}(\sqrt{3}-1)rr_2 - 7)r^2 - 3 r_2^2],$$
and as expected we then have
$$r_2 = \frac{9-4\sqrt{3}}{33} \cdot r \approx 0.063 \cdot r.$$

Now that you should have had fun finding the area of rather-unusual shapes, we should say that these are not newly developed exercises, but rather they go back to 1796, where they were seen among the Japanese *Sangakus'* shapes. The famous French mathematician René Descartes (1595–1650) already tackled this last problem in his four-circle theorem in 1643.[6] These circle problems were rediscovered by the English chemist Frederick Soddy (1877–1956), and so are sometimes referred to as the "Soddy circles" since he published a poem called *The Kiss Precise*.[7]

QUADRILATERAL CURIOSITIES— THE SURPRISING APPEARANCE OF A PARALLELOGRAM[8]

Quadrilaterals often take a backseat to triangles, since most of linear geometry reverts back to triangulation. Even the study of quadrilaterals is done largely through their partition into triangles. Let us now consider some curious properties of the quadrilateral. Suppose you were to draw any shape quadrilateral, and then join (with segments) the midpoints of consecutive sides. What would you expect the resulting quadrilateral to look like? Try drawing a random quadrilateral, preferably one in which no sides have the same length nor are parallel. Then locate the midpoints of each of the sides, and join them as we do in figure 2.24. The resulting shape should be a parallelogram—often called a *Varignon parallelogram*.[9]

In case you think this might have been a coincidence, try drawing another quadrilateral repeating the same process of joining the midpoints of the sides, and once again you will notice that you have just formed another parallelogram. We may wish to conclude that a *quadrilateral formed by joining the midpoints of consecutive sides of any quadrilateral is a parallelogram.*

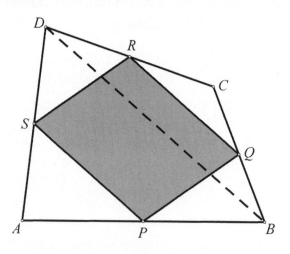

Figure 2.24

This can be easily justified. In figure 2.24, points P, Q, R, and S are the midpoints of the sides of quadrilateral $ABCD$. Draw the diagonal DB. In triangle ABD, PS is a midline[10] and therefore, $PS \parallel BD$, and $PS = \frac{1}{2}BD$. Similarly, in triangle BCD, QR is a midline, and therefore, $QR \parallel BD$, and $QR = \frac{1}{2}BD$. It then follows that $PS \parallel QR$, and $PS = QR$, which establishes that $PQRS$ is a parallelogram. Clearly, $PQ \parallel SR$, and $PQ = SR$.

One might then ask, using this technique of joining the midpoints of the sides of a quadrilateral: what type of quadrilateral $ABCD$ would produce a rectangle *PQRS*, or a rhombus *PQRS*, or a square *PQRS*?

We know that the consecutive sides of a rectangle are perpendicular. Therefore, since the sides of the quadrilateral formed by joining the midpoints of the consecutive sides of a quadrilateral are parallel to the diagonals of the original quadrilateral, it would follow that the sides of the newly formed quadrilateral must be perpendicular, if the original quadrilateral's diagonals are perpendicular. This allows us to conclude that *a quadrilateral formed by joining the midpoints of consecutive sides of a quadrilateral whose diagonals are perpendicular is a rectangle.*

When a parallelogram has all sides equal, it is a rhombus. Therefore, since the sides of the parallelogram formed by joining the midpoints of the sides of a quadrilateral are each half the length of those diagonals, we can conclude that when the diagonals of that quadrilateral are equal, the resulting parallelogram formed by joining the midpoints of the sides of that quadrilateral is a rhombus. In other words, *a quadrilateral formed by joining the midpoints of consecutive sides of a quadrilateral whose diagonals are congruent is a rhombus.* We show this in figure 2.25.

Combining these two facts—namely, if a given rhombus is also shown to be a rectangle—we are well on our way to determining how this quadrilateral formed by joining midpoints of an original quadrilateral would be a square. That can be summarized as follows: *The quadrilateral formed by joining the midpoints of consecutive sides of a quadrilateral whose diagonals are perpendicular and congruent is a square.*

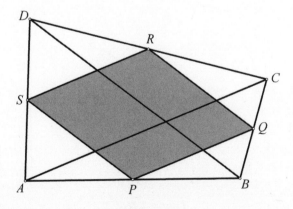

Figure 2.25

MORE QUADRILATERAL CURIOSITIES— THE CENTERS OF A QUADRILATERAL

Before we embark on a search for the centers of a quadrilateral, we should recall that the center of a triangle, called the *centroid*, is determined by the point of intersection of the three medians of a triangle (i.e., the lines joining each vertex with the midpoint of the opposite side). This is the point at which a triangle could balance on a pinpoint.[11] You may wish to try this with a cardboard triangle and carefully locating the centroid having it balance on the point of a pencil.

The quadrilateral actually has two centers. The *centroid* of a quadrilateral is that point on which a quadrilateral of uniform density will balance. This is a point analogous to the centroid of a triangle, where if you have a cardboard quadrilateral and you wish to balance it on the point of a pencil, you would need to locate the centroid. This point may be found in the following way. Let K and L be the centroids of triangle ABC and triangle ACD, respectively. (See figure 2.26.) Let M and N be the centroids of triangle ABD and triangle BCD, respectively. The point of intersection, G, of KL and MN is the centroid of the quadrilateral $ABCD$.

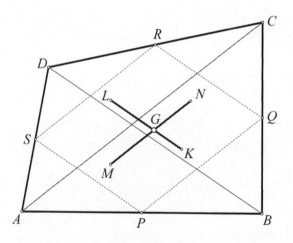

Figure 2.26

There is yet another center of a quadrilateral. This is a point at which you would balance a quadrilateral comprised of only four vertices and no area between them. This could look like four sticks on the same plane with various weights at the vertices. We call this center the *centerpoint* of a quadrilateral, and it is the point of intersection of the two segments joining the midpoints of the opposite sides of the quadrilateral. In figure 2.27, point *H* is the centerpoint of quadrilateral *ABCD*.

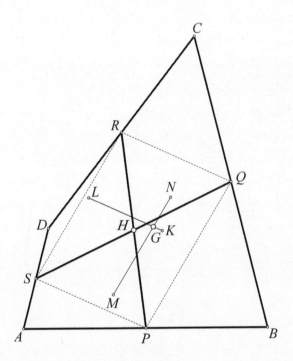

Figure 2.27

If we examine the point of intersection of the lines joining the midpoints of a quadrilateral, the centerpoint, we would find that they bisect each other—that is, they share a common midpoint. This can be easily justified, since these two segments are, in fact, the diagonals of the parallelogram formed by joining the midpoints of the consecutive sides of the quadrilateral, and we know that these diagonals bisect each other. In figure

2.28, points P, Q, R, and S are the midpoints of the sides of quadrilateral $ABCD$. The centerpoint H is determined by the intersection of PR and QS.

While we are in this interesting configuration, and have before us these surprising line-segment relationships, another curious and unexpected property belongs to this centerpoint H. If we consider the respective midpoints, U and V, of the diagonals AC and DB of the original quadrilateral, and draw the line segment UV, we find that the point H bisects UV. This, too, can be easily justified.

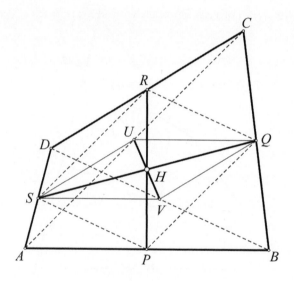

Figure 2.28

In figure 2.28, U is the midpoint of AC, and V is the midpoint of BD. Also Q and S are the midpoints of two of the sides of quadrilateral $ABCD$. In triangle ABC, QU is a midline; therefore, $QU \parallel AB$ and $QU = \frac{1}{2}AB$. Similarly, in triangle ABD, SV is a midline; therefore, $SV \parallel AB$ and $SV = \frac{1}{2}AB$. Thus $QU \parallel SV$, and $QU = SV$. In the same way we have $QV \parallel SU$, and $QV = SU$. It follows that $SVQU$ is a parallelogram. The diagonals of a parallelogram bisect each other, so that UV and QS share a common midpoint, H, which was earlier established as the centerpoint of the quadrilateral.

We can take another step further by showing one more curious relationship about the lengths of the lines we just considered above.

That is, the *sum of the squares of the lengths of the diagonals of any quadrilateral equals twice the sum of the squares of the lengths of the two segments joining the midpoints of the opposite sides of the quadrilateral.* In other words, in figure 2.29, we have

$$AC^2 + BD^2 = 2\,(PR^2 + QS^2).$$

This is a rather-neat connection between the two sets of diagonals—those of the original quadrilateral, and those of the parallelogram formed by joining the consecutive midpoints of the sides of that quadrilateral.

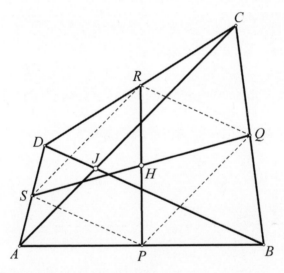

Figure 2.29

For us to justify this relationship we will first have to consider a relatively unknown relationship between the sides and the diagonals of a parallelogram—one that will remind us of the Pythagorean theorem. Namely, that the *sum of the squares of the lengths of the sides of a parallelogram equals the sum of the squares of the lengths of the diagonals.*[12]

With this relationship, we can easily justify the relationship between the diagonals of the two quadrilaterals mentioned above. We established earlier that

$PQ = \frac{1}{2}AC$, and $RS = \frac{1}{2}AC$.

This gives us: $PQ^2 = \frac{1}{4}AC^2$ and $RS^2 = \frac{1}{4}AC^2$.   (1)

Similarly, $QR = \frac{1}{2}BD$, and $PS = \frac{1}{2}BD$.

This gives us: $QR^2 = \frac{1}{4}BD^2$ and $PS^2 = \frac{1}{4}BD^2$.   (2)

Now applying this relationship, just mentioned above, about the sides and the diagonals of a parallelogram, and applying it to parallelogram $PQRS$, we can state:

$$PQ^2 + QR^2 + RS^2 + PS^2 = PR^2 + QS^2.   (3)$$

Making the appropriate substitutions of (1) and (2) into (3) gives us:

$$\frac{1}{4}AC^2 + \frac{1}{4}BD^2 + \frac{1}{4}AC^2 + \frac{1}{4}BD^2 = PR^2 + QS^2$$

$$\frac{1}{2}AC^2 + \frac{1}{2}BD^2 = PR^2 + QS^2$$

$$AC^2 + BD^2 = 2(PR^2 + QS^2).$$

This is what we sought to justify.

Now let's focus our attention first on the general parallelogram, and place squares on the sides of this parallelogram. We show one such *randomly* selected parallelogram $ABCD$, which is in figure 2.30, has squares drawn on its sides. We can locate the centers of each of the squares by getting the point of intersection of each square's diagonals. Joining these four points, amazingly, gives us another square $PQRS$. This theorem is called the *Yaglom-Barlotti theorem* (1955)[13] or is sometimes named the *Thébault theorem* (1937).[14]

When we constructed the squares on each of the sides of the parallelogram, we drew them externally. We could just as well have constructed them internally (that is, so that they overlap the parallelogram) and the result would be the same—their centers would determine another square.

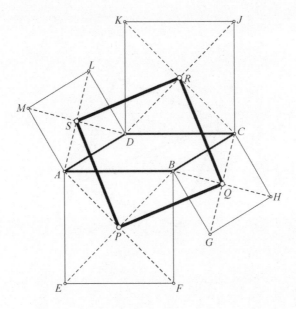

Figure 2.30

Moreover, the diagonals of this square (*PQRS*) meet at the same point (*T*) as the intersection of the diagonals of the original parallelogram (see figure 2.31). To fully appreciate this unexpected relationship, you need to keep in mind that this will hold true for any parallelogram.

We consider the case of the squares erected on the outside of the parallelogram. The triangles *APS*, *BPQ*, *CQR*, and *DRS* are obviously congruent (side-angle-side), so that the quadrilateral *PQRS* is a rhombus. Further, $\angle APS = \angle BPQ$, which implies $\angle QPS = \angle APB = 90°$. Had the squares been drawn on the opposite sides of each of the sides of the parallelogram—namely, overlapping the parallelogram itself—the above proof would still hold true.

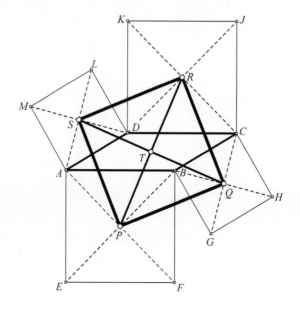

Figure 2.31

We will once again draw squares on the sides of a randomly drawn quadrilateral, as we show in figure 2.32. There you will notice that the line segments joining the centers of the four squares are equal and are perpendicular to each other. This was first published in 1865 by the French engineer Édouard Collignon (1831–1913), but this is today often named after Henri H. van Aubel[15] (1830–1906) as *van Aubel's theorem* (1878). Another way of saying the same thing is that the center points of the four squares form the vertices of an *orthodiagonal quadrilateral*.[16]

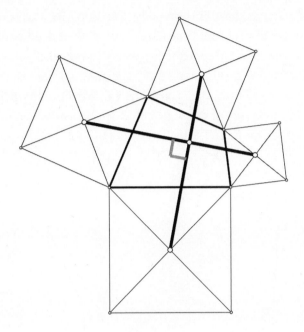

Figure 2.32

It is interesting to see what kind of quadrilateral we would get if we join the centers of the squares drawn (externally) on the sides of various other kinds of quadrilaterals, such as rectangles, squares, trapezoids, and kites.[17]

A NEGLECTED SPECIAL QUADRILATERAL[18]

From the study of high-school geometry, we are made familiar with the common quadrilaterals: the square, the rectangle, the rhombus, the parallelogram, and the trapezoid. What is rarely brought to the attention of students studying elementary geometry is the quadrilateral whose four vertices lie on the same circle. This quadrilateral, often called a *cyclic quadrilateral*, has many interesting properties. It is well known that every triangle is such that its vertices will always lie on a circle. However, this is not the case for quadrilaterals, where only some quadrilaterals, cyclic quadrilaterals, have this property. The opposite angles of a cyclic quadrilateral are always supplementary—that is, their degree sum is 180.

A cyclic quadrilateral sometimes appears unexpectedly. For example, if we construct the angle bisectors of any general quadrilateral, these bisectors will intersect to form a cyclic quadrilateral. We show this in figure 2.33, where the four bisectors of the angles of general quadrilateral $ABCD$ meet to form quadrilateral $EFGH$, which turns out to be a cyclic quadrilateral.

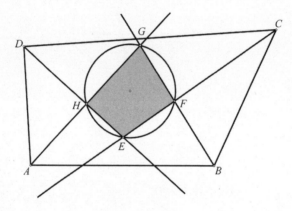

Figure 2.33

To prove that quadrilateral $EFGH$ is cyclic, we will recall that for any quadrilateral (in this case $ABCD$) the sum of the angles

$\angle BAD + \angle ABC + \angle BCD + \angle CDA = 360°$;

therefore, $\frac{1}{2} \angle BAD + \frac{1}{2} \angle ABC + \frac{1}{2} \angle BCD + \frac{1}{2} \angle CDA = \frac{1}{2} \cdot 360° = 180°$.

Substituting, we get: $\angle BAG + \angle ABG + \angle DCE + \angle CDE = 180°$.   (1)

Let us focus on triangle ABG and triangle CDE, where the sum of their angles is:

$(\angle BAG + \angle ABG + \boldsymbol{\angle AGB}) + (DCE + \angle CDE + \boldsymbol{\angle CED}) = 2 \cdot 180°$.   (2)

By subtracting (1) from (2), we find that: $\angle AGB + \angle CED = 180°$.

Since one pair of opposite angles of quadrilateral $EFGH$ are supplementary, the other pair must also be supplementary, and hence quadrilateral *EFGH* is cyclic.

Perhaps one of the most famous theorems involving cyclic quadrilaterals is that attributed to Claudius Ptolemaeus of Alexandria (popularly known as Ptolemy, who lived from about 100 to 180 CE). In his major astronomical work, the *Almagest*[19] (ca. 150 CE), he states this property of cyclic quadrilaterals.

The product of the diagonals of a cyclic quadrilateral equals the sum of the products of the pairs of opposite sides. This is often known as *Ptolemy's theorem.*[20]

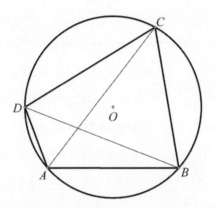

Figure 2.34

We can apply Ptolemy's theorem to the quadrilateral *ABCD* in figure 2.34. There we have $AB \cdot CD + AD \cdot BC = AC \cdot BD$.

Ptolemy's theorem allows us to establish many interesting relationships among the sides and diagonals of a cyclic quadrilateral. For example, we can find the ratio of the diagonals in terms of the side lengths as follows: $\frac{AC}{BD} = \frac{AB \cdot AC + BC \cdot CD}{AB \cdot BC + AD \cdot CD}$. Although it appears somewhat cumbersome, without Ptolemy's theorem, we would be hard-pressed to establish this relationship.[21]

We know that a rectangle is a cyclic quadrilateral. Can you guess which famous theorem can be established if we apply Ptolemy's theorem to a rectangle? You were right if you said the Pythagorean theorem. You can see this in figure 2.35, where rectangle *ABCD* has lengths *a*, widths *b*, and diagonals *c*. Applying Ptolemy's theorem we get: $AB \cdot CD + AD \cdot BC = AC \cdot BD$ or $a \cdot a + b \cdot b = c \cdot c$, which can be written as $a^2 + b^2 = c^2$.

From this we can readily recognize as the Pythagorean theorem applied to triangle *ABC*.

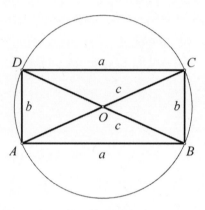

Figure 2.35

SQUARING A SQUARE—A PARTITIONING

The challenge before us now is to determine if a square can be partitioned into a number of noncongruent squares, such that each of the squares has side lengths that are integers and that there is no part of the original square not covered by one of these smaller squares.

There have been several problems in mathematics that have challenged mathematicians for long periods of time. One such problem was that of *squaring the circle*.[22] After about two thousand years and many mathematicians' efforts, in 1882 the German mathematician Ferdinand von Lindemann (1852–1939) put this problem to rest with a proof that it could not be done. In comparison, the question of squaring a square, that is, partitioning a square into a finite number of smaller squares with integral side lengths, first emerged in the last century. In 1998, in honor of the International Congress of Mathematicians (ICM) the German government issued a postage stamp (figure 2.36) that exhibited three specific math-ematical concepts. It showed the value of π expanded to many decimal places. The use of the colors in the stamp is in honor of the solution some years earlier of the famous four-color map problem that was solved in 1976 by Kenneth Appel (1932–2013) and Wolfgang Haken (1928–), who proved, with the help of a computer, that any map can be colored with no more than four colors so that no two territories sharing a common border would be colored with the same color. However, the dominating theme of this stamp, which measures $35 \cdot 35$ millimeters, is figure showing the eleven smaller squares covering the area of the larger square. (See figures 2.36 and 2.37.)

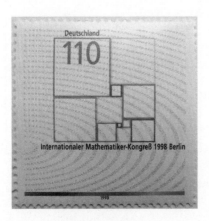

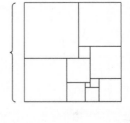

Figure 2.36   **Figure 2.37**

Unfortunately, appearances can be deceptive. The larger quadrilat-eral into which the eleven smaller squares are placed is in actuality not a square, but rather a rectangle. The rectangle on the stamp has dimensions $177 \cdot 176$, while the smallest of the squares has dimensions that are 9×9. This tiling was first discovered by the British mathematician Arthur Harold Stone (1916–2000), when in 1940 he wanted to prove that a perfect square partitioning of a square was not possible.[23] (See figure 2.38.)

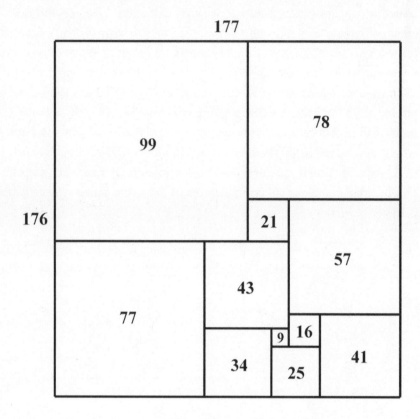

Figure 2.38

We notice that the rectangle is almost a square, but in fact, is not a square. We might call this an "almost square"—one in which the length and width differ by one unit.

We might ask a question, is there such an "almost square" that can be partitioned into fewer than eleven smaller squares? The answer is yes, since a rectangle, which we can consider an almost square with dimensions $33 \cdot 32$ can be partitioned into nine smaller squares whose side lengths are as follows: 1, 4, 7, 8, 9, 10, 14, 15, and 18 (figure 2.39). This partition of the almost square was first accomplished in 1925 by Polish mathematician Zbigniew Moroń (1904–1971).[24]

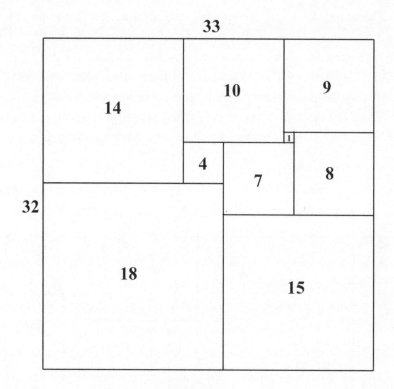

Figure 2.39

In 1940, H. Reichardt and H. Toepken proved that the squaring dissection of a rectangle required at least nine smaller squares.[25] In fact, there are only two rectangles that permit partitioning into nine noncongruent squares and there are six rectangles that can be partitioned in this fashion by ten noncongruent squares, and Moroń discovered that the partitioning of a rectangle with dimensions 65 × 47 would require ten smaller noncongruent squares. Figure 2.40 summarizes the number of rectangles (*a*) that can be partitioned into *n* smaller noncongruent squares.

n	9	10	11	12	13	14	15	16	17	18
a	2	6	22	67	213	744	2,609	9,016	31,426	110,381

Figure 2.40

The search for a perfect squaring of a square—namely to be able to partition a square into smaller noncongruent squares was motivated by the 1907 book *The Canterbury Puzzles* by Henry Ernest Dudeney (1857–1930). This attempt provided by one of the problems in the book also had a slight omission preventing it from properly solving the problem.

There are basically three kinds of "squaring of the square." The first one we can look at is where all the smaller squares are different sizes, as shown in figure 2.41.

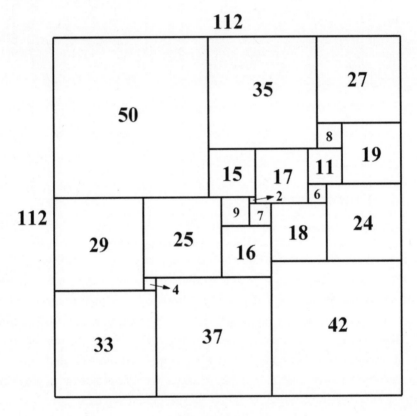

Figure 2.41

The second kind of "squaring of a square" is where not all of the smaller squares are different, as shown in figure 2.42. Here there are two squares each of dimensions 1, 2, 3, 5, and 11.

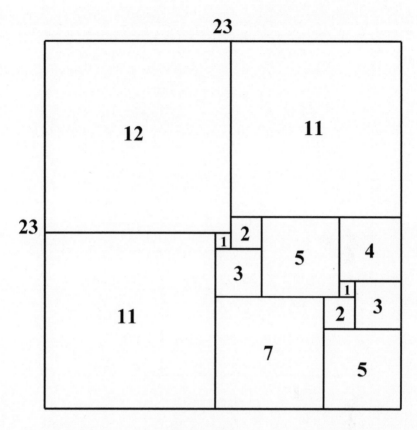

Figure 2.42

The third kind of "squaring of the square" is where a subportion of the square is in fact a rectangle which is also partitioned exactly to smaller squares, as shown in figure 2.43. The squares are of dimensions: 1, 2, 3, 4, 5, 8, 9, 14, 16, 18, 20, 29, 30, 31, 33, 35, 38, 39, 43, 51, 55, 56, 64, and 81. The 111 × 94 rectangle has subsquares of dimensions: 1, 3, 4, 5, 9, 14, 16, 18, 20, 38, 39, 55, and 56.

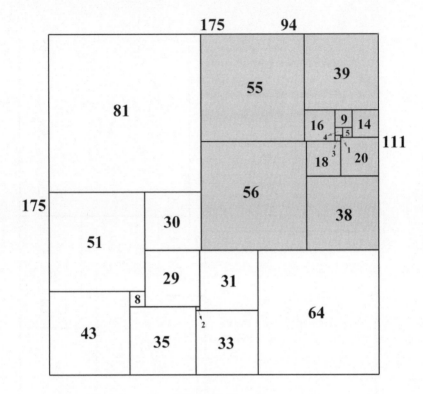

Figure 2.43

Finally, in 1964 J. C. Wilson discovered a perfect squaring of the square using twenty-five noncongruent smaller squares to cover the area of a square of dimensions 503 × 503. The smallest perfect squaring of the square was accomplished by the Dutch mathematician Adrianus Johannes Wilhelmus Duijvestijn (1927–1998), who, with the aid of computer, discovered a square of side length 112, which she was able to partition into twenty-one smaller squares of different dimensions: 2, 4, 6, 7, 8, 9, 11, 15, 16, 17, 18, 19, 24, 25, 27, 29, 33, 35, 37, 42, and 50. This can be seen by the calculation of areas:

$$2^2 + 4^2 + 6^2 + 7^2 + 8^2 + 9^2 + 11^2 + 15^2 + 16^2 + 17^2 + 18^2 + 19^2 + 24^2 + 25^2 + 27^2 + 29^2 + 33^2 + 35^2 + 37^2 + 42^2 + 50^2 = 12,544 = 112^2.$$

In 1989/1990, C. Müller and J. D. Skinner proved[26] that for all natural numbers $n \geq 21$, a perfect squaring of the square exists. The number of squares (a) for the n partitioning with squares is summarized in figure 2.44.

n	21	22	23	24	25	26	27	28	29
a	1	8	12	26	160	441	1 152	3 001	7 901

Figure 2.44

Our present-day mathematicians are still fascinated with squaring squares both with different noncongruent squares as well as with a repetition of smaller partitioning squares.

A CURIOUS PLACEMENT OF TRIANGLES[27]

In the study of geometry we seem to be most concerned about two triangles being congruent—having the same shape and same area. However, two triangles can also be related by their position on the plane. For example, consider two triangles, ABC and $A'B'C'$ (of possibly different shapes), whose corresponding sides (extended) meet in three collinear points X, Y, and Z (i.e., points that lie on the same straight line) and have:

sides AC and $A'C'$ meet at point X,
sides BC and $B'C'$ meet at point Y, and
sides AB and $A'B'$ meet at point Z,

then the lines joining the corresponding vertices (AA', BB', and CC') are concurrent (in point P), as shown in figure 2.45. This famous two-triangle relationship was first discovered by the French mathematician and engineer Gérard Desargues (1591–1661) and today bears his name. By the way, the converse of this relationship is also true. That is, if two triangles are so placed that the lines joining their corresponding vertices are concurrent (in figure 2.45, point P is that point of concurrency), then the extensions of their corresponding sides will meet in three collinear points (points X, Y, and Z).[28]

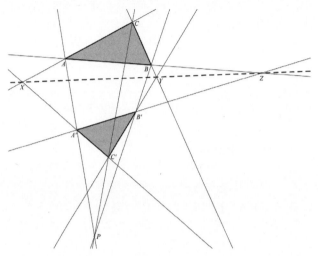

Figure 2.45

KEPLER'S CONJECTURE

Amazingly, after more than four hundred years, in 1998, the famous Kepler conjecture has been proved. What exactly is the Kepler conjecture? Before we answer the question, let us consider the problem that a grocer has when stacking his oranges in a way that would minimize the amount of space required. Ideally, the optimum stacking requires that beginning at the bottom row, each succeeding row has the oranges shifted by half a sphere, as shown in figures 2.46 and 2.47—essentially forming a hexagonal figure.

Figure 2.46. (Image courtesy of Ingo Henze.)   **Figure 2.47. (Image from View FotoCommunity, by Boizo.)**

It is quite certain that a grocer is unaware that it was not until 1998 that this sort of stacking was first proved to be the ideal when attempting to minimize display space. This was Kepler's conjecture—namely, that this sort of stacking used the least amount of space—as opposed to another sort of stacking shown in figure 2.46.

Sir Walter Raleigh (ca.1554–1618) asked his assistant, the mathematician Thomas Harriot (1560–1621), how many cannonballs can be stacked in a wagon in order to maximize the number of cannonballs into the given space. (See figure 2.48.) In 1606, this question was taken to the famous mathematician and astronomer Johannes Kepler (1571–1630). In 1611, Kepler conjectured that the optimal packing was achieved when the cannonballs were stacked in such a way that each ball was placed so that it touched exactly twelve other balls.[29]

**Figure 2.48. Cannonballs in the
Arsenal of Metz during the Siege of Paris in 1871.**

The German mathematician Carl Friedrich Gauss (1777–1855) discovered that the grocer's regular stacking was the most efficient, as it filled 74.05 percent of the space available ($\frac{\pi}{\sqrt{18}} = 0.7404804896\ldots$).

The possibility still existed that some irregular stacking could save more space. In 1900, the German mathematician David Hilbert (1862–1943) challenged the mathematics community to come up with the ideal packing arrangement as problem 18 of his famous list of twenty-three most important unsolved problems in mathematics.

The analogous problem on the plane is relatively simple. There, the optimum circle packing would be considered hexagonal packing and covers more than 90 percent of the plane area. (See figure 2.49.) More precisely we have $\frac{\pi}{\sqrt{12}} = 0.9068996821\ldots$:

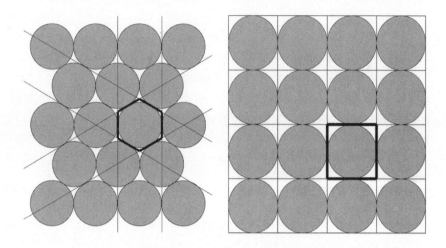

Figure 2.49. Hexagonal packing: **Figure 2.50. Square packing:**
$\frac{\pi}{\sqrt{12}} = 0.90689\ldots$ $\frac{\pi}{4} = 0.78539\ldots$

This is only true when we consider covering an infinite plane. On a finite plane for $n = 1, 4, 9, 16$, and 25 circles, the square packing (figure 2.50) is optimal.[30]

In 1998 the American mathematician Thomas C. Hales (1958–) proved that in three-dimensional space the most compact packing of spheres is either a cubic packing or the hexagonal packing mentioned earlier. Each of these has an average density of $\frac{\pi}{\sqrt{18}}$.

Hales provided the results of his computer application of this problem, which should have quelled the long-standing disagreement among mathematicians on this topic of sphere packing. As early as 1990, Wu-Yi Hsiang (1937–) developed a proof of Kepler's conjecture that was constantly being corrected through 1993, when it was published in the *International Journal of Mathematics*. Although some mathematicians still attacked the proof, Wu-Yi Hsiang did not retract it. There is no further discussion about the veracity of Hales's proof.

Just as the two-dimensional situation is rather simple, we find that the three-dimensional analog is quite difficult. The Kepler conjecture has been one of the longest standing unsolved problems in the history of mathematics. We seem now to be comfortable with it.

Geometry offers practically endless curiosities, but the limitation of this chapter had us present those more "off the beaten path." One of the more popular curiosities has to do with showing proofs of geometric paradoxes, such as "proving" that all triangles are isosceles. We recommend the book *Magnificent Mistakes in Mathematics*,[31] which has lots of such demonstrations.

BIBLIOGRAPHY

Fukagawa, Hidetoshi, and Dan Pedoe. *Japanese Temple Geometry Problems*. Winnipeg: The Charles Babbage Research Center, 1989.

Fukagawa, Hidetoshi, and Dan Pedoe. *How to Resolve Japanese Temple Geometry Problems*. Tōkyō: Mori Kitashuppan, 1991.

Fukagawa, Hidetosh, and Tony Rothman. *Sacred Mathematics: Japanese Temple Geometry*. Princeton, NJ: Princeton University Press, 2008.

Huvent, Géry. *Sangaku: Le mystère des énigmes géométriques japonaises*. Paris: Dunod, 2008.

Posamentier, Alfred S. *Advanced Euclidean Geometry*. Hoboken, NJ: John Wiley & Sons, 1999.

Rothman, Tony, and Hidetoshi Fukagawa. (1998–2005). "Japanese Temple Geometry," *Scientific American*, May 1998, pp. 84–91.

Scriba, Christoph J., and Peter Schreiber. *5000 Jahre Geometrie: Geschichte, Kulturen, Menschen*. Heidelberg: Springer-Verlag, 2010.

Bogomolny, Alexander. "Interactive Mathematical Miscellany and Puzzles," http://www.cut-the-knot.org/pythagoras/Sangaku.shtml (accessed March 28, 2014).

Bogomolny, Alexander. "Malfatti's Problem," http://www.cut-the-knot.org/Curriculum/Geometry/Malfatti.shtml (accessed March 28, 2014).

Chapter 3

CURIOUS PROBLEMS
WITH CURIOUS
SOLUTIONS

In this chapter we will present a wide variety of mathematical problems—many are somewhat "off the beaten path"—that will lend themselves quite interestingly to some rather-curious solutions. Aside from being entertaining—sometimes due to their nature, they will also provide the reader with some gee-whiz, surprisingly simple—yet often-overlooked—solutions. These problems should also present an enlightening message to the reader about alternative approaches to solving problems—both simple and complicated. To make this part of the book more interesting, we have decided to separate the problems from their solutions, since oftentimes when reading a problem the temptation to look a bit farther to the solution is so great that it spoils the fun of the initial struggle with the problem. As they say in athletics: "no pain, no gain" also may hold true in mathematics problem solving. The beauty will lie in the very clever and simple solutions. We frequently give a bit of attention to traditional approaches to attacking problems, and then show how, with some clever thinking, a trivial solution—often overlooked—brings a level of enlightenment that we hope will make these carefully selected problems entertaining.

Before we begin our journey through a collection of these problems, we would like to alert you to some of the pitfalls that can be encountered in mathematical problems in everyday life. For example, in planning for the installation of an air-conditioning system, sheet-metal ductwork had to be installed. The installation required the ductwork to turn a right-angle corner through a very narrow passageway. The contractor had suggested that in order to turn the corner he would use the same amount of sheet metal as he did to make the $8'' \cdot 8''$ ducts throughout, but to turn the corner in the narrow space, he would change the dimensions to $4'' \cdot 12''$, thus using the same amount of sheet metal. Was this a correct solution to the problem of turning the corner? Clearly, the answer is no! The contractor was perplexed when he was told that this solution was unacceptable. He didn't realize that the cross section of the original ductwork had an area of $8'' \cdot 8'' = 64$ in², while the cross-section of the narrower ductwork only had a cross-section area of $4'' \cdot 12'' = 48$-in², which would have a great deal of difference on the air supply. Only when it was explained with an extreme example—that is considering a duct of $\frac{1}{2}'' \cdot 31''$ did he realize the error of his suggestion. (He obviously did not get the contract for the installation!)

There are problems that appear simple, and where the solution seems impossible to find. However, with traditional methods these problems can be solved, but when we consider a non-traditional solution, we are led to amazement by the simplicity of the solution that seemed to have passed us by. Such problems define a curiosity in mathematics that adds to its beauty! In that spirit, we offer our first problem.

THE PROBLEMS

PROBLEM 1

We begin with a problem that seems to be counterintuitive.

Would a bathtub drain as quickly with a single drain hole of diameter two inches, as with two drain holes of diameter one inch?

PROBLEM 2

The calendar always presents interesting challenges—perhaps because of its peculiar structure.

What is the probability that the following dates fall on the same day of the week: April 4, June 6, August 8, October 10, and December 12?

PROBLEM 3

With this problem we are testing the reader's concentration and logical thinking.

What time is it now when in two hours it will be half as long till noon as in one hour from now?

Another time-related question is the following: What time is it now, when there is twice as much of the day left as has already passed? (Here we consider a twenty-four-hour day beginning at midnight.)

PROBLEM 4

Having mastered the previous problem, you might find this more challenging.

If the day before yesterday was two days after Monday, which day in the week is the day after tomorrow?

If you got that one, now try this question: Which day the week is today, when the day before yesterday till Wednesday is twice as many days as from yesterday till tomorrow?

PROBLEM 5

The water lilies in a pond double every day. After one hundred days, the pond is completely covered with water lilies. How many days did it take for the pond to be half-covered?

We continue our journey through curious problems with a few simple arithmetic problems, where you can appreciate the solution, since it is quite unexpected and rather simple.

PROBLEM 6

Given the four numbers

$$7,895$$
$$13,127$$
$$51,873$$
$$7,356$$

what percent of their sum is their average?

PROBLEM 7

Which is greater, 25 percent of $76 or 76 percent of $25?

PROBLEM 8

We have 100 kg of berries and water, where 99 percent of the weight is water. A while later, the water content of the mixture is 98 percent water. How much do the berries weigh?

PROBLEM 9

What is the smallest number that is divisible by each of the nine digits in our base-10 number system?

PROBLEM 10

This problem requires the reader to look at "the bigger picture" and not be distracted by details.

In figure 3.1 we have a square and an isosceles right triangle with the right-angle vertex at the center of the square. We have $CE = 2$, $CF = 6$, and $AB = 8$. What part of the square is the shaded quadrilateral?

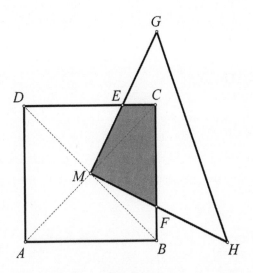

Figure 3.1

PROBLEM 11

The shape in figure 3.2 can be partitioned into four congruent shapes as shown in figure 3.3. Show how this figure can be partitioned into five congruent shapes.

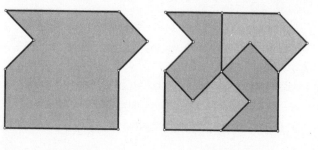

Figure 3.2   Figure 3.3

PROBLEM 12

This tests your logical thinking!

A standard deck of fifty-two playing cards is randomly split into two piles with twenty-six cards in each. How does the number of red cards in one pile compare to the number of black cards in the other pile?

PROBLEM 13

Here we present a problem that requires nothing more than some simple elementary algebra and that should rapidly lead to a correct solution. However, even here a clever alternate solution will be provided after we go through the traditional approach.

We are given that $\frac{1}{x+5} = 4$. What is the value of $\frac{1}{x+6}$?

PROBLEM 14

A problem that has vexed many mathematics recreationalists is the following question: Maria is twenty-four years old. Maria is twice as old as Anna was when Maria was as old as Anna is now. How old is Anna?

PROBLEM 15

An even more challenging problem with ages is now presented to exhibit the curious ways of mathematical thinking.

When Max is twice as old as he is now, the sum of the ages of Max and Jack will be forty-eight. Then Max will be ten years older as Jack was when he was twice as old as Max was when he was one-third as old as Jack will be in fifteen years. What are the current ages of Max and Jack?

PROBLEM 16

This problem might remind the reader of the kind of verbal problem found in most algebra textbooks; however, there is a deceiving point to this that wouldn't be found in one of those typical uniform-motion problems.

There are two trains, one going from Chicago to New York and the other from New York to Chicago, which is a distance of eight hundred miles. One is traveling uniformly at sixty miles per hour, and the other is traveling at forty miles per hour. They start toward each other, at the same time, along the same tracks. At the same time, a bee begins to fly from the front of one of the trains, at a speed of eighty miles per hour, toward the oncoming train. After touching the front of this second train, the bee reverses direction and flies toward the first train (still at the same speed of eighty miles per hour). The bee continues this back-and-forth flying until the two trains collide, crushing the bee. How many miles did the bee fly?

We will now consider a number of problems that have to deal with weighing items on a balance scale. These problems require some logical thinking, and some clever approaches. In each case, the cleverness of the approach should be impressive.

PROBLEM 17

Consider a bag with nine coins that are all identical in appearance, however, one of the coins is lighter than the other eight coins. With only two weighings on a balance scale, how can we determine which of the nine coins is a lighter one?

PROBLEM 18

Coin-weighing problems seem to tweak a certain logical-thinking procedure. In that spirit, another is provided here.

In a bag of four coins that all look the same, two coins of equal weight are heavier than the other two coins, which are also of equal weight. With only two weighings on a balance scale, how can we determine which are the two heavier coins?

PROBLEM 19

This problem about weighings takes a different tack.

Using the balance scale, we would like to weigh objects weighing from one pound to thirteen pounds—not fractional weights. How can this be done with only three different weights? And what should these three weights be?

PROBLEM 20

Each of ten court jewelers gave the king's advisor, Mr. Pogner, a stack of gold coins. Each stack contained ten coins. The real coins weighed exactly one ounce each. However, one and only one stack contained "light" coins, each having had exactly 0.1 ounce of gold shaved off the edge. Mr. Pogner wishes to identify the crooked jeweler and the stack of light coins with just one single weighing on a balance scale. How can he do this?

PROBLEM 21

Now that we have had practice with weighings, consider this more challenging problem.

Suppose you have twelve coins that all look exactly the same. However, one of the coins is defective and has a different weight from the other eleven coins. How can we determine the defective coin using a balance scale and only three weighings?

PROBLEM 22

Some problems involving infinity can be very "upsetting." Infinity presents some unusual ways of thinking, as we have seen in chapter 1. Now consider the next few problems that will show the unusual way of thinking when it comes to considering infinity.

Here is a rather curious-looking problem where a proper path to a solution may at first be a bit daunting. We are asked to find the value of x that satisfies the equation:

$$x^{x^{x^{x^{x^{x^{\cdots}}}}}} = 2.$$

PROBLEM 23

A strategy similar to that used in problem 22 can be used to find x in the following equation:

$$x = \sqrt{2\sqrt{2\sqrt{2\sqrt{2\sqrt{2\sqrt{2\sqrt{2\sqrt{2\sqrt{2\sqrt{2}}}}}}}}}}\,\cdots$$

PROBLEM 24

A young boy was given a basket of eggs after having worked in an egg farm for the day. By accident, a man driving a tractor hits the basket and breaks the eggs. The driver offers to pay the young boy for the broken eggs, and asks him how many eggs he had in the basket. The boy does not remember the exact number, but he remembers having been told that if you take the eggs two at a time, there would be one egg remaining. He further remembered that if you took the eggs in groups of three, four, five, and so on until you took them ten at a time, there would be also one less than the number taken remaining each time. However, if you took them eleven at a time, there would be none remaining. What is the smallest number of eggs that the boy could have had in the basket?

PROBLEM 25

What is the smallest number with exactly twenty-eight divisors?

PROBLEM 26

Find the smallest prime number consisting of ten digits, where all digits are different, and, of course, the first digit cannot be zero.

PROBLEM 27

How can we show that the number $z = 3^n + 63$ is divisible by 72 when n is an even number?

PROBLEM 28

In figure 3.4 we have nine wheels touching each other with diameters successively increasing by 1 cm. Beginning with 1 cm as the smallest circle, and 9 cm for the largest circle, how many degrees does the largest circle turn when the smallest circle turns by 90°?

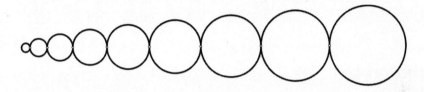

Figure 3.4

PROBLEM 29

Here is a rather harmless-looking problem that could be offered to any beginning elementary algebra student.

We are given the following facts: the sum of two numbers is 2, and the product of the same two numbers is 3. We are asked to find the sum of the reciprocals of these two numbers.

PROBLEM 30

Show that the equation $x^4 - 5x^3 - 4x^2 - 7x + 4 = 0$ has no negative roots.

PROBLEM 31

Simplify each of the following without a calculator:

(a) $\dfrac{729^{35} - 81^{52}}{27^{69}}$,

(b) $\dfrac{6 \cdot 27^{12} + 2 \cdot 81^{9}}{8000000^{2}} \cdot \dfrac{80 \cdot 32^{3} \cdot 125^{4}}{9^{19} - 729^{6}}$.

PROBLEM 32

Here is a problem where a surprising solution makes it trivial!

In a single-elimination basketball tournament with twenty-five teams competing, how many games must be played in order to get a winner?

PROBLEM 33

The nifty solution of the previous problem should make the following problem less confusing.

At a tennis club, thirty-two players enter a single-elimination tournament. Two of the players are Mr. Wagner and Mr. Strauss. What are the chances that these two players will be playing each other in this tournament?

PROBLEM 34

To extend the amount of wine in a sixteen-ounce bottle, David decides on the following procedure. On the first day, he will drink only one ounce of the wine and then refill the bottle with water. On the second day, he will drink two ounces of the water-wine mixture, and then again refill the bottle with water. On the third day, he will drink three ounces of the water-wine mixture, and again refill the bottle with water. He will continue this procedure for succeeding days until he empties the bottle by drinking sixteen ounces of the mixture on the sixteenth day. How many ounces of water will David drink altogether?

PROBLEM 35

One member of the RW Society, which has one hundred members, has been notified that the society's meeting place must be changed. This member activates the society's telephone squad by telephoning three other members, each of whom then telephones another three members, and so on, until all one hundred RW Society members have been notified of the meeting-place change. What is the greatest number of members of the RW Society who do not need to make a telephone call?

PROBLEM 36

There are problems that can be solved in a very curious and unexpected fashion. One such is the following, which seems quite harmless and yet can be quite challenging if a certain approach is not used.

We have two one-liter bottles. One contains a half-liter of red wine and the other contains a half-liter of white wine. We take a tablespoonful of red wine and pour it into the white-wine bottle. Then we take a tablespoon of this new mixture (white wine and red wine) and pour it into the bottle of red wine. Is there more red wine in the white-wine bottle, or more white wine in the red-wine bottle?

PROBLEM 37

Here we are given five circles tangent to one another and placed as shown in figure 3.5. The problem here is to find a line through the center point of the left-most circle that will equally divide the areas of these five circles.

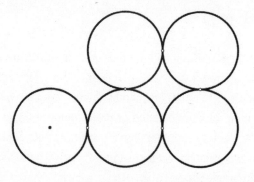

Figure 3.5

PROBLEM 38

We are given here three congruent squares with a side length of 1. We are asked to find the sum of the angles $\alpha + \beta + \gamma$, as shown in figure 3.6.

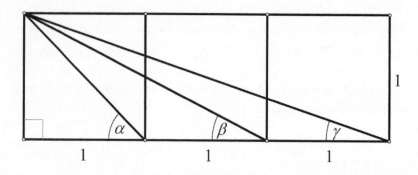

Figure 3.6

PROBLEM 39

In figure 3.7 we have a semicircle with the point P randomly placed on the diameter. Points A and B are situated on the circle such that they form angles of 60° with the diameter as shown in the figure. This problem asks us to show that the length of AB is equal to the radius of the semicircle.

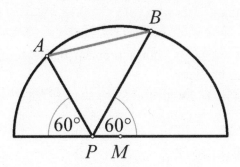

Figure 3.7

PROBLEM 40

This problem is one that appears to be dauntingly impossible to grapple with. However, follow along, after trying to solve it, and appreciate the unexpected solution.

If the sum of the divisors of 360 is 1,170, then what is the sum of the reciprocals of these divisors?

PROBLEM 41

Here we are faced with another seemingly overwhelming task, but watch the solution that makes this simple!

Find the units digit of the number equal to

(a) 8^{19} and
(b) 7^{197}.

(Naturally, we this should be done without a calculator or a computer.)

PROBLEM 42

Once again we are faced with a problem that asks us to work with inordinately large numbers.

Find the units digit for the following sum: $13^{25} + 4^{81} + 5^{411}$.

PROBLEM 43

Here is another problem that will lead us to a clever "ah-ha" solution.

What is the quotient of 1 divided by 500,000,000,000?

This can be restated as, "find the value of $\frac{1}{500,000,000,000}$."

PROBLEM 44

What is the sum of $1^3 + 2^3 + 3^3 + 4^3 + \ldots + 9^3 + 10^3$?

PROBLEM 45

There are entertainments in mathematics that stretch (gently, of course) the mind in a very pleasant and satisfying way. Here is one such problem.

Where on earth can you be so that when you can walk: *one mile south*, then *one mile west*, and then *one mile north*, you end up at the starting point?

PROBLEM 46

Sometimes, giving an airplane pilot flying directions could be misleading. Suppose an aircraft carrier were placed in the Gulf of Guinea, at the point where the prime meridian and the equator intersect. A plane takes off exactly in a northeastern direction and remains on that course continuously. The question is, where will that plane end up?

PROBLEM 47

In figure 3.8 we have two equilateral triangles, one is inscribed in a circle, and the other one is circumscribed about the same circle. We need to find the ratio of the areas of the two equilateral triangles and the ratio of their sides.

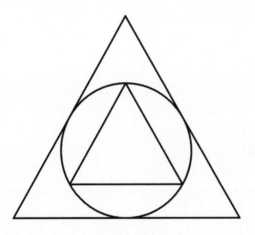

Figure 3.8

PROBLEM 48

A rectangle whose sides are in the ratio of $1:\sqrt{5}$ is inscribed in a circle whose radius is 6 units long. (See figure 3.9.) By joining the midpoints of the sides of the rectangle, a rhombus inscribed in the rectangle is formed. What are the lengths of the sides of this rhombus?

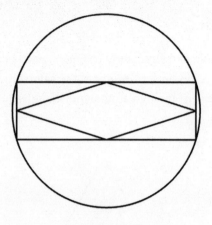

Figure 3.9

PROBLEM 49

Here we have a circle inscribed in a square and a second square inscribed in the circle, as shown in figure 3.10. Find the ratio of the areas of the two squares.

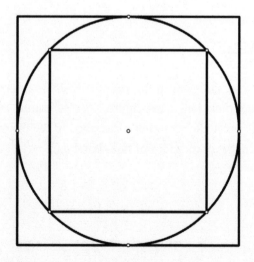

Figure 3.10

PROBLEM 50

The simplicity of this problem is well hidden, but when we are shown its simplicity, we tend to be annoyed with ourselves for not having seen it immediately. In figure 3.11, a quarter circle, arc *FCE*, is drawn with center at point *A*. From a randomly selected point *C* on the quarter arc perpendiculars are drawn to *AF* and *AE*. If the length of *AE* is 10, what is the length of *BD*?

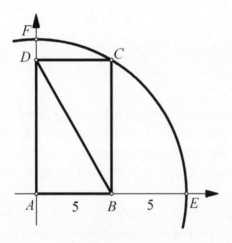

Figure 3.11

PROBLEM 51

There are problems in mathematics that play on our logical thinking. That is to say, the problem abuses our natural course of thinking. We offer one such problem here. Given nine dots as arranged in figure 3.12, without lifting the pencil, use four straight lines to connect all nine dots.

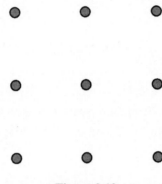

Figure 3.12

PROBLEM 52

Sometimes changing your thinking pattern, as was the case in the previous problem, can be quite useful. The following problem once again requests some open thinking—logical reasoning.[1]

In the configuration shown in figure 3.13, there are eleven sticks in each outside row and column.

Show how to remove one stick from each row, and one from each column, and still remain with eleven sticks in each row and column.

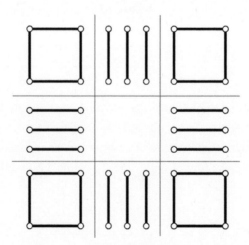

Figure 3.13

PROBLEM 53

While thinking of minimal movements, as we did previously, consider the following problem.

Suppose there is a lineup of six glasses where three are empty and three are full of water, as shown in figure 3.14. By picking up only one glass, what can you do to change the arrangement so that no empty glass is next to another empty glass, and no full glass is next to another full glass?

Figure 3.14

PROBLEM 54

Sam was given four pieces of chain, each consisting of three links (see figure 3.15). Show how these four pieces of chain could be made into a circular chain by opening and closing, *at most*, three links.

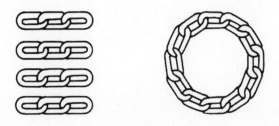

Figure 3.15

PROBLEM 55

There are times when a problem looks deceptively simple to solve and yet seems to evade a solution. The following is one such problem.

In figure 3.16, there are two parallelograms that share a common vertex, and where the vertex of each one lies on the side of the other. That is, vertex *G* is on side *CD*, vertex *B* is on side *EF*, and point *A* is the common point of the vertices of the two parallelograms. The problem here is to determine the ratio of the areas of the two parallelograms, *ABCD* and *AEFG*.

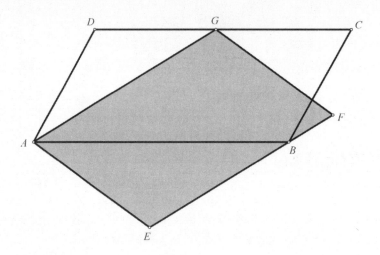

Figure 3.16

PROBLEM 56

The key to the solution of this problem is to find a symmetric part of the figure given.

Consider the ten dots arranged in the shape of an equilateral triangle shown in figure 3.17. Show how, by moving only three dots, the orientation (or direction) of the triangle can be reversed.

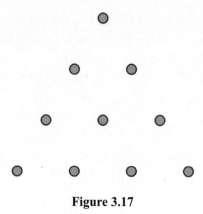

Figure 3.17

PROBLEM 57

A monkey is trying to climb out of a one-hundred-foot-deep well. His efforts to climb out of the well allow him to progress one foot per day as follows: the first half of the day he climbs three feet, and the second half of the day he falls back two feet. How many days will it take the monkey to climb out of the well?

PROBLEM 58

We now have a problem that initially confuses most, yet, by systematically working through the problem, the solution can be found rather easily.

If, on average, a hen and a half can lay an egg and a half in a day and a half, how many eggs at this rate would six hens lay in eight days.

PROBLEM 59

Max discovers that his car radiator is in need of seven liters of water. (See figure 3.18.) He turns off the road next to a stream and notices that in the trunk of his car he has only an eleven-liter can and a five-liter can. His problem is to determine how he can go to the stream with these two cans and measure off exactly seven liters that he would then be able to bring back to the car and pour into his car radiator.

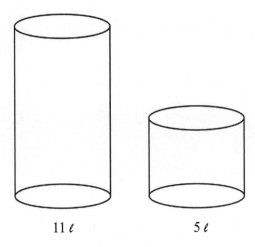

11 ℓ   5 ℓ

Figure 3.18

PROBLEM 60

We typically hear of the concept of relativity with regard to physics. Here we present a simple mathematics problem that uses this concept.

While rowing his boat upstream, Simon drops a cork overboard and continues rowing for ten more minutes. He then turns around, chasing the cork, and retrieves it when the cork has traveled one mile downstream. What is the rate of the stream?

PROBLEM 61

This problem presents a real conundrum that will give some deeper insight to the concept of probability. It is one of the more curious aspects of geometric probability. Prepare yourself for an interesting journey in this field of mathematics.

Consider two concentric circles (figure 3.19), where the radius of the smaller circle is one-half that of the larger circle. What is the probability that a point selected randomly in the larger circle is also in the smaller one?

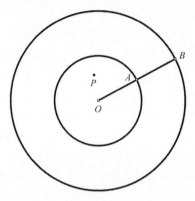

Figure 3.19

PROBLEM 62

In figure 3.20, *AB* is a chord of the larger of the two concentric circles and is tangent to the smaller circle at point *T*. Find the area of the lighter-shaded region, which is between the two circles, if *AB* = 8.

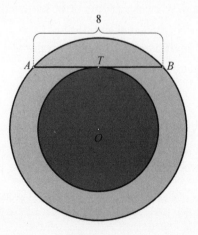

Figure 3.20

196 MATHEMATICAL CURIOSITIES

PROBLEM 63

At first glance, it would appear that there is not enough information given to solve this problem.

There are times when simple arithmetic is not enough to answer a question. These are times when logical reasoning must be used to buttress the arithmetic. A case in point can be seen from this neat little problem.

A woman and her three daughters pass her neighbor's house, and he asks her how old her three daughters are. She responds that, coincidentally, the product of their ages is thirty-six, and the sum of their ages is the same number as his address. He looks puzzled as he stares at the house number and finds no solution, but then he gets even more puzzled when the woman tells him that she almost forgot one essential piece of information: her oldest daughter's name is Miriam. This really baffles him. How can this man determine the ages of the woman's daughters? (We are only dealing with integer ages.)

PROBLEM 64

Although trigonometry is not always in the memory reserve of the average person, this problem may awaken some prior knowledge.

In triangle *ABC*, we find that the product: cos α · cos β · cos γ > 0. What kind of the triangle would have this property? A right triangle? An acute triangle? Or an obtuse triangle?

PROBLEM 65

As we set up this problem, we form a triangle with the natural numbers arranged as follows:

```
                    1
                2   3   4
            5   6   7   8   9
        10  11  12  13  14  15  16
    17  18  19  20  21  22  23  24  25
26  27  28
```

We are asked to find the row in which the number 2000 can be found, and which numbers lead that row and end that row.

PROBLEM 66

Here is a problem that sounds overwhelming but can be made simple with elementary algebra and just a bit of manipulation.

Consider two joggers who begin from their hometown at sunrise. The joggers are going in opposite directions along the same path toward each other's hometown. Running at a uniform pace, they meet at noon. They keep running at the same pace, with the first runner reaching the second runner's hometown at 4 p.m., and the second runner reaching the first runner's hometown at 9 p.m. What time was sunrise on that day?

PROBLEM 67

Too often we take geometry for granted. Here is a problem that puts to good use some of the common principles.

In figure 3.21, we have P as any point on side AB of triangle ABC, where M is the midpoint of BC and N is the midpoint of AC. What is the ratio of the area of quadrilateral $MCNP$ to the area of triangle ABC? (See figure 3.21.)

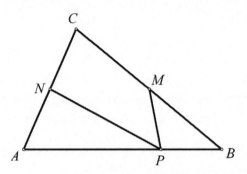

Figure 3.21

PROBLEM 68

Three-dimensional geometry—which is how we view the world in which we live—also presents some unexpected surprises, as seen with the following problem.

We are given a cube with diagonals drawn on each of the faces as shown in figure 3.22. What is the volume of the regular tetrahedron formed by these diagonals, when the side of the cube is length a?

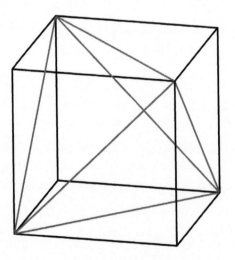

Figure 3.22

PROBLEM 69

There is a question that most people—including mathematicians—have trouble accepting; namely, that a solid as being described below can exist.

We are given the three shapes shown in figure 3.23, where we have a square with a side of length 1, a circle with a diameter of length 1, and an isosceles triangle whose base and altitude have length 1.

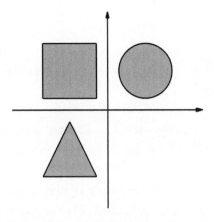

Figure 3.23

Is there a solid figure that, viewed from three different perspectives, has these three shapes? If so, how might it be shaped?

PROBLEM 70

We now embark on one of the most counterintuitive problems in probability. This is a fine example of how we can't always trust our intuition, and that mathematics is our best guide.

What do you think the chances (or probability) are of two classmates having the same birth date (month and day only) in a class of about thirty students?

PROBLEM 71

Here is a curious problem that can be solved either in the traditionally algebraic fashion or by taking a step back and looking at the "bigger picture." Those who reach for a calculator ruin the beauty of the problem.

Which is larger, $\sqrt[9]{9!}$ or $\sqrt[10]{10!}$?

PROBLEM 72

A regular dodecagon is inscribed in a unit circle. A point, P, on the circle is selected. Find the sum of the squares of the distances from P to each of the vertices. (See figure 3.24.)

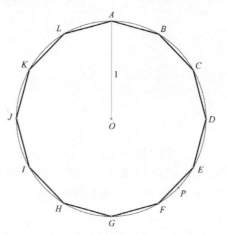

Figure 3.24

PROBLEM 73

When the diagonals are drawn in a regular hexagon as shown in figure 3.25, a second hexagon is formed (also, because of symmetry, it is a regular hexagon) within the first hexagon (shaded in the diagram). What fraction of the area of the larger hexagon is the area of the smaller hexagon?

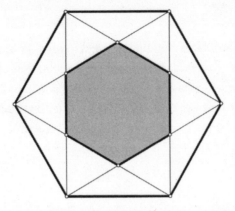

Figure 3.25

PROBLEM 74

In isosceles triangle *ABC*, shown in figure 3.26, the altitude and the base have the same length, that is, *CD* = *AB*. Also, *BE* is perpendicular to *AC*. In this situation, something very unusual occurs, namely the triangle *CEB* turns out to be our well-known 3, 4, 5 right triangle. But how can we show that this is actually true? That is the challenge of this problem.

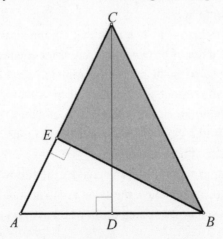

Figure 3.26

PROBLEM 75

There are some problems that appear rather solvable but look as though they require a lot of work. Oftentimes in these situations one needs to step back from the requested task and see if there is a pattern that can be found that can simplify the work considerably. We offer one such problem here:

Find the sum of the terms in the following series:

$$20^2 - 19^2 + 18^2 - 17^2 + 16^2 - 15^2 + \ldots + 4^2 - 3^2 + 2^2 - 1^2.$$

PROBLEM 76

We now offer a problem that requires logical thinking rather than arithmetic!

In a drawer, there are eight blue socks, six green socks, and twelve black socks. What is the smallest number of socks that must be taken from the drawer, without looking at them, to be certain of having two black socks?

PROBLEM 77

More logical thinking here!

In a dark room there is a shoe rack that contains twelve shoes. There are three identical pairs of brown shoes and three identical pairs of black shoes. They were placed on the rack in a completely random order. Because the room is dark, you cannot see the color of the shoes. All you can determine is whether they are right-foot or left-foot shoes. How many shoes must be taken from the closet to be certain that you have either a pair of black shoes or a pair of brown shoes?

Using a similar principle, we can consider the following problem: Are there two New Yorkers who have the same number of hair strands on their head? (Obviously, we are excluding those who have no hair on their head.)

PROBLEM 78

This problem can be solved with a calculator, but it is much more elegant (and efficient) to look for a pattern.

Find the sum of the series $\frac{1}{2} + \frac{1}{6} + \frac{1}{12} + \frac{1}{20} + \frac{1}{30} + \ldots + \frac{1}{2450}$.

PROBLEM 79

This problem can be solved with a calculator or a *computer algebra system*, (CAS), that is, a software program that allows computation over mathematical expressions in a way that is similar to the traditional manual computations of mathematicians and scientists. But, again, it is much more elegant (and efficient) to look for a pattern.

Prove the inequality $\frac{1}{2} \cdot \frac{3}{4} \cdot \frac{5}{6} \cdot \frac{7}{8} \cdot \ldots \cdot \frac{99}{100} < \frac{1}{10}$.

PROBLEM 80

Don't be fooled by this problem; it may be more complicated than it seems.

At exactly what time after four o'clock will the hands of the clock overlap precisely?

PROBLEM 81

Taking the clock as a field for mathematics problems a step further, we offer this problem.

It is clear that at twelve o'clock all three hands of a clock, the second hand, the minute hand, and the hour hand, overlap. How often over the next twelve hours will all three hands overlap?

PROBLEM 82

Here is a simple one, if you see the right path to the answer.

A rectangular box has faces that measure 165 square inches, 176 square inches, and 540 square inches. What is the volume of the box?

PROBLEM 83

For which value of x would this expression be true?

$$\frac{1}{4} \left\{ \frac{1}{4} \left[\frac{1}{4} \left(\frac{1}{4} x - \frac{1}{4} \right) - \frac{1}{4} \right] - \frac{1}{4} \right\} - \frac{1}{4} = 0$$

PROBLEM 84

For readers who remember some trigonometry, we offer the following problem.

Find the value of T, where $T = \tan 15° \cdot \tan 30° \cdot \tan 45° \cdot \tan 60° \cdot \tan 75°$.

PROBLEM 85

Figure 3.27 shows a regular octagon with an isosceles triangle using one of the sides as a base and the vertex angle on the opposite side. What fraction of the area of the octagon is the area of the shaded isosceles triangle?

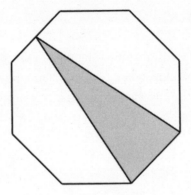

Figure 3.27

Figure 3.28 shows a regular octagon with a rectangle formed by a pair of opposite sides of the octagon. What fraction of the area of the octagon is the area of the shaded rectangle?

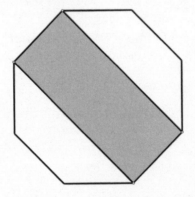

Figure 3.28

PROBLEM 86

Here is a curiously confusing sequence of numbers that is not easily identifiable. What is the next number?

1, 2, 3, 4, 6, 8, 9, 12, 16, 18, 24, 27, 32, 36, 48, 54, 64, 72, 81, 96, . . .

PROBLEM 87

Given a chessboard and thirty-two dominos, each the exact size of two of the squares on the chessboard, can you show how thirty-one of these dominos can cover the chessboard, when a pair of opposite squares have been removed? (See figure 3.29.)

Figure 3.29

PROBLEM 88

The challenging problem we offer here might leave the reader a bit frustrated. Consider the following: there are four people in a cave that has only one exit. Unfortunately, these four people lost their belongings in the cave and had only one flashlight between them. In order to get out of the cave, they need to adhere to the following requirements:

1. The four people vary in their time to reach the exit. Person A requires five hours, person B requires four hours, person C requires two hours, and person D requires one hour.[2]

2. Because the cave is dark and dangerous, only one can walk with the flashlight. For that reason, the flashlight must be passed on from a person reaching the exit to someone still in the cave.
3. The cave is so narrow that, at most, two people at a time can use the path to the exit.
4. There is only one flashlight, and its battery life is exactly twelve hours.

Hint: Mathematicians have been frustrated by this question, as they believe it requires thirteen hours, which would make it impossible for them to leave the cave with a flashlight having a battery life of only twelve hours. There is a solution with twelve hours!

PROBLEM 89

There are at least two ways of doing this problem: The poet's way or the peasant's way. Which will you select?

What is the sum of all the numbers formed by rearranging the digits of the number 975?

PROBLEM 90

A simple algorithm—but well camouflaged!

You can present this problem to friends, but you stand a possibility of losing the friendships because of the frustration that could develop in seeking a solution. Through hints along the way, you could even drive them mad.

The algorithm we seek assigns a natural number to any four-digit number as shown by the examples below. Your friends can select a four-digit number, and you (as the expert) can apply the algorithm to assign the appropriate natural number to this given four-digit number. By doing this so rapidly, it will clearly impress your friends but will also, while they seek a solution, add to their frustration.

The chart below is an example of the algorithm in action, assigning to each of the four-digit numbers a simple natural number from 0 to 9.

1254 → 0	4110 → 1	1378 → 2	3365 → 1
5678 → 3	6780 → 4	1000 → 3	3196 → 2
2266 → 2	4444 → 0	4371 → 0	5358 → 2
3934 → 1	8888 → 8	1378 → 2	8698 → 6
2381 → 2	9699 → 4	1379 → 1	7778 → 2

Have you discovered the algorithm yet? Perhaps the following would help to discover the algorithm.

1000 → 3
1001 → 2
1100 → 2
1110 → 1
1111 → 0

If you want to be nasty, you might mention that an elementary-school child who can count to ten could stumble on the algorithm. The following might be helpful, or perhaps even more frustrating!

6000 → 4
6006 → 4
6600 → 4
6660 → 4
6666 → 4

THE SOLUTIONS

SOLUTION FOR PROBLEM 1

We can approach this problem two ways: arithmetically or geometrically. An arithmetic solution would require us to find the area of each cross-section. The cross-section of the drain hole whose diameter is two inches, and whose radius is one inch, has an area of $\pi(1)^2 = \pi$. The cross-section area of one of the one-inch-diameter drain holes is

$$\pi\left(\frac{1}{2}\right)^2 = \frac{\pi}{4}.$$

Therefore, the sum of the two one-inch-diameter drain holes is $\frac{\pi}{2}$, or half the cross-section area of the larger drain hole.

We can also make the comparison geometrically by merely inspecting the cross-sections superimposed over one another as shown in figure 3.30. Here you can clearly notice that there is a significant difference between the sum of the areas of the two smaller circles and the larger circle.

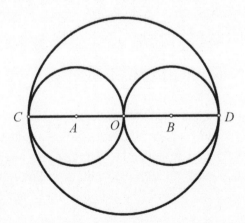

Figure 3.30

SOLUTION FOR PROBLEM 2

When we write these dates numerically, we will notice that the dates chosen are equivalent to the number of the month. That is, the dates in question are: 4/4, 6/6, 8/8, 10/10, and 12/12.

(We are, of course, assuming that the Gregorian calendar will continue indefinitely.)

The answer to the original question will probably baffle most readers, since the answer is that the probability that these dates will all fall on the same day of the week is 1, that is, certainty!

Every year, April 4 is followed *exactly* nine weeks later by June 6, eighteen weeks later by August 8, twenty-seven weeks later by October 10, and thirty-six weeks later by December 12.

This explains why these dates always will fall on the same day of the week.

Another question with a surprising answer is, what is the probability that January 1 of the beginning of a millennium (e.g., January 1, 2001) will land on a Friday, a Saturday, or a Sunday?

Surprisingly, the answer is that the probability is 0, that is, it will never happen. In fact, January 1 at the beginning of a millennium will always be on a Monday or on a Thursday. As a case in point, January 1, 2001, was on a Monday.

SOLUTION FOR PROBLEM 3

Most people would be confused by this question. However, a clever way to approach this curious problem is to work backward. The answer is 9:00 a.m. If it is now 9:00 a.m., it will be 11:00 a.m. in two hours, which is one hour before noon. In one hour after 9:00 a.m., it will be 10:00 a.m., or two hours to noon. Therefore, from 11:00 a.m. to noon is half as long as from 10:00 a.m. to noon.

We will use some simple algebra to make this second question more manageable. Let x represent the number of hours that have already passed since midnight. Since $24 - x$ is the amount of time remaining in the day, and that is twice as long as x, then, $2x = 24 - x$. Solving this equation, we get $3x = 24$, and $x = 8$. Therefore, at eight o'clock we have satisfied question's requirements.

SOLUTION FOR PROBLEM 4

Once one gets past the verbiage, the problem is not so difficult. If the day before yesterday was a Monday, then two days later is a Wednesday. Therefore, today is Friday, and the day after tomorrow would be Sunday. We just have to make sure not to get lost in the setup of the problem.

For the second question, working backward would be desirable, since the number of days from yesterday until tomorrow is two days. Twice as many days would, of course, be four days. If four days past the day before yesterday gives us Wednesday, then we count backward four days from Wednesday to get to Saturday. Finally, if Saturday is the day before yesterday, then today must be Monday.

No less confusing would be the question, if yesterday had been Wednesday's tomorrow and tomorrow is Sunday's yesterday, what day would today be? See if you can arrive at the right answer: Friday!

SOLUTION FOR PROBLEM 5

This is a classic problem that tricks many unsuspecting readers. Once again, working backward is a good strategy toward a quick solution. Since we know that the number of water lilies doubles on a daily basis, and on the hundredth day the pond is fully covered, then on the day before it had to be half-covered. Therefore, it required ninety-nine days to cover half the pond.

SOLUTION FOR PROBLEM 6

When faced with this question, we typically do exactly what is being called for. That is, first we find the sum of the four numbers; then we find their average. Finally, we divide and convert to required percent:

$$7{,}895 + 13{,}127 + 51{,}873 + 7{,}356 = 80{,}251 \text{ and}$$

$$\frac{80251}{4} = 20{,}062.75 \text{ and } \frac{20{,}062.75}{80251} = \frac{1}{4} = 0.25, \text{ that is, } 25 \text{ percent.}$$

However, it might be wise here, and perhaps simpler, to first consider the general case. We shall let the sum of these numbers be represented by S. Then their average is $\frac{S}{4}$.

To find what percent the average is of the sum, we first divide $\frac{\frac{S}{4}}{S} = \frac{1}{4}$.

We now change $\frac{1}{4}$ to a percent—which the problem calls for—to get 25 percent. We have, therefore, avoided a great deal of unnecessary calculation by simply stepping back from the problem and considering the general case and observing how it presents us with the answer.

What may be surprising is that the result is independent of the four given numbers!

SOLUTION FOR PROBLEM 7

The answer is that they are both the same: $19.

In this case, what appears to be a curious problem is merely a situation where we should not be bogged down with the calculation, rather we should notice that there is a commutativity here, since in both cases we are multiplying the "same two numbers."

In one case we have:

$\frac{25}{100} \cdot \$76 = \frac{25 \cdot 76}{100}$ (= $19); and in the other we find that $\frac{76}{100} \cdot \$25 = \frac{76 \cdot 25}{100}$ (= $19).

SOLUTION FOR PROBLEM 8

A common wrong answer is that with an evaporation of 1 percent water, 99 percent must be the berries, which would imply that the berries weigh 99 kg. This is wrong!

Initially, the berries are 99 percent water, meaning that they contain 99 kg of water and 1 kg of dry matter, or 1 percent of the berries' mass. The mass of the dry matter doesn't change: at the end of the drying process, its weight remains 1 kg. In the meantime, however, the proportion of the total mass that is not water has doubled, to 2 percent.

In order for something that has a fixed quantity (our 1 kg of dry material) to double in proportion (going from 1 percent to 2 percent), the total amount of stuff has to be cut in half. We began with 1 percent or 1/100 dry, and ended with 2 percent or 2/100 dry, which reduces to 1/50—meaning we end with one kilogram of dry matter out of fifty kilograms total. Thus, we have 49 kg of water at the end.

SOLUTION FOR PROBLEM 9

Naturally, one could multiply the numbers from 1 to 9 and get 362,880, which is a rather-large number and clearly not the smallest number that is divisible by these digits. The question, then, is how can we find the smallest such number? Someone experienced with mathematics would say that the number would be the product of the numbers 5, 7, 8, and 9, which is 2,520. But then one may wonder how the other numbers, which have not been listed here, have been accounted for as divisors of 2,520.

Let's take a look at these remaining numbers. The number 1 is clearly a trivial case. Since we included 8 among the first group, we have indirectly included the numbers 2 and 4 as well. Within the number 9 we certainly included the number 3. Since 2 and 3 have been included their product, the number 6, is also now included as a divisor. In this way, we have included all the numbers from 1 to 9 as divisors by merely taking the product $5 \cdot 7 \cdot 8 \cdot 9 = 2,520$, making it the smallest number divisible by the numbers 1 through 9.

SOLUTION FOR PROBLEM 10

The answer to this question is that the quadrilateral's area is one quarter of the area of the square. This is independent of the nature of the right triangle MGH just as long as it is right-angle vertex is at the center, M, of the square. (Perhaps some of the given information was distracting—intentionally so—namely, that $CE = 2$, $CF = 6$, and $AB = 8$.) We can justify that very easily, by noting that $\triangle MEC \cong \triangle MFB$. (See figure 3.31.) It is clear that the area of triangle BMC is one quarter of the area of a square. Therefore, we simply replace the two congruent triangles to give us the conclusion that the area of the quadrilateral $ECFM$ is also then one-quarter the area of the square.

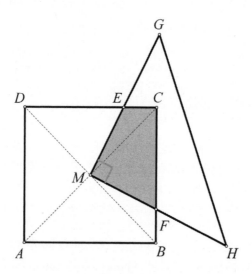

Figure 3.31

SOLUTION FOR PROBLEM 11

We are easily distracted, and perhaps misled, by the complicated partitioning provided in the statement of the problem. The solution is deceptively simple, as shown in figure 3.32.

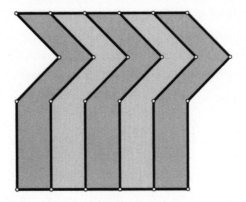

Figure 3.32

SOLUTION FOR PROBLEM 12

The typical way that one approaches this problem is to represent the number of black cards and the number of red cards in each pile in a symbolic fashion. We can represent the situation symbolically as follows:

$B1$ = the number of black cards in pile #1
$B2$ = the number of black cards in pile #2
$R1$ = the number of red cards in pile #1
$R2$ = the number of red cards in pile #2

Then, since the total number of black cards equals twenty-six, we can write this as $B1 + B2 = 26$, and since the total number of cards in pile #2 equals twenty-six, we have $R2 + B2 = 26$.

By subtracting these two equations: $B1 + B2 = 26$, and $R2 + B2 = 26$, we get: $B1 - R2 = 0$. Therefore, we have $B1 = R2$, which tells us that the number of red cards in one pile equals the number of black cards in the other pile. Although this solves the problem, there is nothing elegant about the solution. Our theme in this chapter is to provide clever solutions to show the beauty and power in mathematics.

As an alternate, and perhaps as a more clever approach, we shall take all the red cards in pile #1 and switch them with the black cards in pile #2. Now, all the black cards will be in one pile, and all the red cards will be in pile # 2. Therefore, the number of red cards in one pile and the number of black cards in the other pile had to be equal to begin with. Simple logic solves the problem!

SOLUTION FOR PROBLEM 13

The traditional solution is simply to solve the given equation, $\frac{1}{x+5}=4$, and find the value of x, which it turns out is $x=-\frac{19}{4}$. We then substitute for x in the fractional expression $\frac{1}{x+6}$ and obtain $\frac{4}{5}$. Of course, this may involve a bit of some cumbersome algebraic and arithmetic manipulations, but it is certainly correct.

Perhaps a more clever way to approach this problem is to begin with the given information: the equation $\frac{1}{x+5}=4$ = 4. If we take the reciprocals of both sides of the equation to obtain $x + 5 = \frac{1}{4}$, we will have something far more manageable. Since we are looking for $x + 6$, we merely have to add 1 to both sides of this equation to get $x+5+1=\frac{1}{4}+1$, or $x + 6 = \frac{5}{4}$. We again take the reciprocals of both sides to obtain $\frac{1}{x+6}=\frac{4}{5}$, which is what we were asked to find. Clearly this might be considered a more elegant approach—or at least a more curious one.

SOLUTION FOR PROBLEM 14

The solution to this problem does not lend itself to simply setting up an equation that will lead us to an answer. More is involved. We can begin by setting up the following table (figure 3.33):

	previously	currently
Anna	a	$a + x$
Maria	$24 - x$	24

Figure 3.33

We have $24 = 2a$, therefore, $a = 12$. Also $24 - x = a + x = 12 + x$; therefore, $x = 6$.

Anna was twelve when Maria was as old (eighteen) as Anna is currently (eighteen).

Alternatively, we could also have proceeded as follows:

The situation presented is manifested at two levels:

1. at the current time, when Maria is twenty-four years old, and
2. at a time n years ago.

We then set up the following relationships: M = Maria's age ($= 24$), A = Anna's age, and n = the difference between the two time periods.

From the first part: Maria is twice as old as Anna was:

$$2 (A - n) = M \tag{1}$$

From the second part: when Maria was as old as Anna is now:

$$M - n = A \tag{2}$$

Equation (2) is now substituted into equation (1)

$$2 (M - n - n) = M \Rightarrow n = \frac{M}{4} = \frac{24}{4} = 6 \tag{3}$$

The value of $n = 6$ inserted into equation (2), yields:

$$M - 6 = A \Rightarrow A = 24 - 6 = 18 \tag{4}$$

This tells us that Anna is eighteen years old.

SOLUTION FOR PROBLEM 15

Let M represent Max's present age and J represent Jack's present age. When Max will be twice as old as he is currently, Jack will be $J + M$ years old. Thus, the first sentence of the problem gives us the following equation: $2M + (J + M) = 48$.

Simplified that is, $3M + J = 48$.   (1)

In fifteen years, Jack will be $J + 15$ years old, and one-third of that age is $\frac{1}{3}J + 5$.

Twice that age is $\frac{2}{3}J + 10$. Ten years beyond that age is $\frac{2}{3}J + 20$.

The second part of the problem can be stated as: $2M = \frac{2}{3}J + 20$, which can be simplified as:

$3M - J = 30$.   (2)

When we add equations (1) and (2), we get $6M = 78$, and $M = 13$.

Then, from equation (1), we have $J = 48 - 3M = 48 - 39 = 9$.

That means that Max is currently thirteen years old, and Jack is currently nine years old.

SOLUTION FOR PROBLEM 16

One is naturally drawn to find the individual distances that the bee traveled. An immediate reaction is to set up an equation based on the familiar relationship: "rate times time equals distance." However, this back-and-forth path is rather difficult to determine, requiring considerable calculation. Even then, it is very difficult to solve the problem in this fashion.

A much more elegant approach would be to solve a simpler analogous problem (one might also say we can look at the problem from a different point of view). We seek to find the distance the bee traveled. If we knew the time the bee traveled, we could determine the bee's distance because we already know the speed at which the bee is flying.

The time the bee traveled can be easily calculated, since it traveled the entire time the two trains were traveling—until they collided. To determine the time the trains traveled, t, we set up an equation as follows:

The distance of the first train is $60t$, and the distance of the second train is $40t$. The total distance the two trains traveled is 800 miles. Therefore, $60t + 40t = 800$, and $t = 8$ hours, which is also the time the bee traveled. We can now find the distance the bee traveled, which is $8 \cdot 80 = 640$ miles. What seemed to be an incredibly difficult task, that of finding the distance the bee traveled back and forth, has been reduced to a rather-simple application of a common "uniform-motion problem" of the sort encountered in the elementary algebra course with the solution readily apparent.

SOLUTION FOR PROBLEM 17

At first reading, this may seem like an impossible task. However, if we take three coins at random and then take a second batch of three coins and place them on either side of the balance scale, we find that if the two batches of three coins weigh the same, then clearly the lighter coin is among the three coins not yet weighed. In this case, we take two of the coins that have not been weighed, and place one coin on each end of the balance scale. If they balance, then we know that the remaining third coin is the lighter one. If they don't balance, the lighter coin is clearly identified on the balance scale.

If, on the other hand, on the first weighing of the two batches of three randomly selected coins, the two batches do not balance, then we can determine in which batch of three coins the lighter coin is a member. We then proceed as we did previously with the third batch (above) of three coins to determine which is the lightest of the nine coins.

SOLUTION FOR PROBLEM 18

We begin by taking two coins and placing them on either end of the balance scale. If the two coins balance, then we simply take one of these coins and weigh it against one of the not-yet-weighed coins. This will immediately tell us which are the heavier two coins, as the representative coins from each of the two groupings will determine. On the other hand, if the initial two coins do not balance, then we will know which is the heavier of the two coins. And then repeating this procedure for the two not-yet-weighed coins will reveal which of those two coins is the heavier.

SOLUTION FOR PROBLEM 19

The three required weights must be: a one-pound weight, a three-pound weight, and a nine-pound weight. For example, to weigh an object that weighs two pounds, we would use the one-pound weight and the three-pound weight as follows: We place the three-pound weight and the one-pound weight on either side of the balance scale. The required pound weight (to be weighed) would then be placed alongside the one-pound weight, which would then be balanced and determine the item of two pounds.

Of course, the three-pound weight on one side of the scale would suffice to weigh three pounds on the other side of the scale. To weigh an item of four pounds, we simply place the one-pound and the three-pound weights on one side of the scale, with the other side of the scale determining four pounds. The remaining weighings are shown below:

five pounds: 9 lb. on one side of the scale and 3 + 1 lb. (along with the 5 lb.) on the other side of the scale

six pounds: 9 lb. on one side of the scale and 3 lb. (along with the 6 lb.) on the other side of the scale

seven pounds: 9 + 1 lb. on one side of the scale and 3 lb. (along with the 7 lb.) on the other side of the scale

eight pounds: 9 lb. on one side of the scale and 1 lb. (along with the 8 lb.) on the other side of the scale

nine pounds: 9 lb. on the other side of the scale

ten pounds: 9 + 1 lb. on the other side of the scale

eleven pounds: 9 + 3 lb. on one side of the scale and 1 lb. (along with the 11 lb.) on the other side of the scale

twelve pounds: 9 + 3 lb. on the other side of the scale

thirteen pounds: 9 + 3 + 1 lb. on the other side of the scale

SOLUTION FOR PROBLEM 20

The traditional procedure is to begin by selecting one of the stacks at random and weighing it. This trial-and-error technique offers only a one-in-ten chance of being correct. Once this is recognized, one may revert to attempt to solve the problem by reasoning. First of all, if all the coins were true, their total weight would be $10 \cdot 10 = 100$ ounces. Each of the ten counterfeit coins is lighter, so there will be a deficiency of $10 \cdot 0.1 = 1$ ounce. But thinking in terms of the overall deficiency doesn't lead anywhere, since the one-ounce shortage will occur whether the counterfeit coins are in the first stack, the second stack, the third stack, and so on.

Let us try to solve the problem by organizing the data in a different fashion. We must find a method for varying the deficiency in a way that permits us to identify the stack from which the counterfeit coins are taken. Label the stacks #1, #2, #3, #4, . . . #9, #10. Then we take one coin from stack #1, two coins from stack #2, three coins from stack #3, four coins from stack #4, and so on. We now have a total of $1 + 2 + 3 + 4 + . . . + 8 + 9 + 10 = 55$ coins. If they were all true, the total weight would be 55 ounces. If the deficiency were 0.5 ounces, then there were five light coins, taken from stack #5. If the deficiency were 0.7 ounces, then there were seven light coins, taken from stack #7, and so on. Thus, Mr. Pogner could readily identify the stack of light coins and consequently the jeweler who had shaved each coin.

SOLUTION FOR PROBLEM 21

To accomplish this task we take any four coins and weigh them against any other four coins. If the scale balances, then we know that the defective coin is not among these eight coins, but rather among the remaining four coins.

In this case, we will weigh three of the not-yet-weighed coins against three of those coins that have already been weighed. If the scale balances, then we have identified the defective coin as the remaining coin, which had not yet been weighed. Additional weighing would then be required to determine if the defective coin is heavier or lighter than the other coins. This could be done by simply placing this defective coin on the scale against one of the other coins.

If, on the other hand, the above weighing of three coins versus three coins does not balance, we are well on our way to determining that the defective coin is heavier or lighter than the rest. We take the lighter side of the balance scale (the side where the scale is up) and weigh two of the coins. The lighter of the two coins will then be the defective coin. However, if the two coins weigh the same, then the third coin is the defective and lighter coin. An analogous procedure can be followed to determine if the coin is heavier than the rest.

Let us go back to the first weighing. This time, suppose that our original weighing of four coins against another four coins does not balance. This would immediately eliminate from any future weighings the remaining four coins that were not weighed at this time. This weighing also determines that one set of four coins may contain the heavier defective coin, or that the other set of four coins contains a lighter defective coin. To make matters clearer at this point, we will designate the four coins that we suspect might have a heavier defective coin among them as H-coins and those in the set that may contain a lighter defective coin as L-coins. We will consider the other four coins, which we know have the normal weight as N-coins.

Our next weighing will then have three H-coins and one L-coin on one side of the balance scale, and one H-coin and three N-coins on the other side of the balance scale, while leaving three L-coins and one N-coin off the scale. If the scale balances, then one of the three L-coins, which were left off the scale, is defective. We then use the previous procedure to determine the defectively light coin. That is, we weigh two of the coins. The lighter of the two coins will then be the defective coin. However, if the two coins weigh the same, then the third coin is the defective and lighter coin.

If, on other hand, the scale does not balance when weighing three H-coins and one L-coin on one side of the balance scale, and one H-coin and three N-coins on the other side of the balance scale, then there are two possibilities: the side with three H-coins and one L-coin is either heavier or lighter than the other side. Let us take these two possibilities one at a time.

If the side with three H-coins and one L-coin is heavier, that would indicate that one of the three H-coins on that side of the scale is defective and heavier than the others. We then weigh two of the H-coins. The heavier of the two H-coins will then be the defective coin. However, if the two H-coins weigh the same, then the third H-coin is the defective and heavier coin.

If the side with three H-coins and one L-coin is lighter, that would indicate that either the L-coin on that side of the scale is defectively lighter than the other coins, or the H-coin on the other side of the scale is defectively heavier than the other coins. The defective coin can be determined by weighing either of these suspected defective coins against a normal N-coin. (See figure 3.34.[3])

Now that you have survived this rather-lengthy solution, you can pretty much feel like you may have gone through the gamut of these coin-weighing problems. These show another dimension of the mathematics problems that require no arithmetic—just logical thinking!

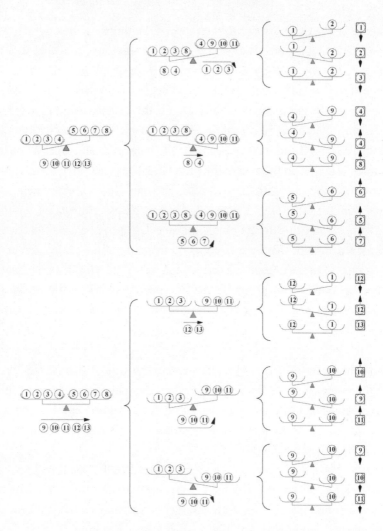

Figure 3.34

SOLUTION FOR PROBLEM 22

At first glance, most people would be overwhelmed and not know how to approach the problem. However, we offer here a rather-simple solution to a very complicated-looking equation. We begin by noticing that there is an infinite number of x's in this series of powers. Eliminating one of the

x's would not have any effect on the end result, because of the nature of infinity. Therefore, by removing the first x, we find that all the remaining x's must also equal 2. This then permits us to rewrite this equation as $x^2 = 2$. It then follows that $x = \pm\sqrt{2}$. If we remain in the set of positive real numbers, then the answer is $x = \sqrt{2}$.

Below you can see how the successive increases get ever closer to 2.

$$\sqrt{2} = 1.414213562\ldots$$

$$\sqrt{2}^{\sqrt{2}} = 1.632526919\ldots$$

$$\sqrt{2}^{\sqrt{2}^{\sqrt{2}}} = 1.760839555\ldots$$

$$\sqrt{2}^{\sqrt{2}^{\sqrt{2}^{\sqrt{2}}}} = 1.840910869\ldots$$

$$\sqrt{2}^{\sqrt{2}^{\sqrt{2}^{\sqrt{2}^{\sqrt{2}}}}} = 1.892712696\ldots$$

$$\sqrt{2}^{\sqrt{2}^{\sqrt{2}^{\sqrt{2}^{\sqrt{2}^{\sqrt{2}}}}}} = 1.926999701\ldots$$

$\ldots$

And so we have a surprisingly simple solution to a very complicated-looking problem.

SOLUTION FOR PROBLEM 23

Once again we have an infinite number of radicals, and so we can square both sides of this equation to get:

$$x^2 = 2\sqrt{2\sqrt{2\sqrt{2\sqrt{2\sqrt{2\sqrt{2\sqrt{2\sqrt{2\sqrt{2\sqrt{2\sqrt{2}}}}}}}}}}}\ldots$$

However, the expression under the radical, which has an infinite number of such radicals, is simply equal to x, since that is what we were given. We can, therefore, change the last equation to read $x^2 = 2x$. We know that $x \neq 0$, therefore the only other possibility is that $x = 2$. Now you may be wondering, how can x equal 2 when x equals this nest of radicals? What we have here is that as the number of these radicals approaches infinity, they get closer and closer to the value 2. You can try this with a calculator and you will notice how the value of x gets closer and closer to 2 with each additional $\sqrt{2}$.

$$\sqrt{2} = 1.414213562\ldots$$

$$\sqrt{2\sqrt{2}} = 1.681792830\ldots$$

$$\sqrt{2\sqrt{2\sqrt{2}}} = 1.834008086\ldots$$

$$\sqrt{2\sqrt{2\sqrt{2\sqrt{2}}}} = 1.915206561\ldots$$

$$\sqrt{2\sqrt{2\sqrt{2\sqrt{2\sqrt{2}}}}} = 1.957144124\ldots$$

$$\sqrt{2\sqrt{2\sqrt{2\sqrt{2\sqrt{2\sqrt{2}}}}}} = 1.978456026\ldots$$

. . .

This is one of the curiosities that occurs when we deal with infinity.

SOLUTION FOR PROBLEM 24

This problem requires searching for a number—and then the smallest number—that has the following characteristics:

when it is divided by 10, it leaves a remainder of 9;
when it is divided by 9, it leaves a remainder of 8;
when it is divided by 8, it leaves a remainder of 7;
when it is divided by 7, it leaves a remainder of 6;
. . . and so on . . .
when it is divided by 3, it leaves remainder of 2;
and when it is divided by 2, it leaves a remainder of 1.

One such number is 3,628,799. But what is the smallest such number? The lowest common multiple of the numbers 1, 2, 3, 4, 5, 6, 7, 8, 9, 10 is $2 \cdot 2 \cdot 2 \cdot 3 \cdot 3 \cdot 5 \cdot 7 = 2{,}520$. Therefore, 2,519 is the smallest number that will leave these remainders.

SOLUTION FOR PROBLEM 25

We know that $28 = 2 \cdot 2 \cdot 7$. The number we seek must be of the form $2^p \cdot 3^r \cdot 5^s$ (by using the first three prime numbers), where $(p + 1) \cdot (r + 1) \cdot (s + 1) = 28$. Therefore, $p = 6$, $r = 1$, and $s = 1$, so that we get $2^6 \cdot 3^1 \cdot 5^1 = 960$. The question looked complicated, but it becomes curiously simpler when you see a logical approach!

SOLUTION FOR PROBLEM 26

Recall that if, and only if, the sum of the digits of any number is divisible by 3, then the number itself is divisible by 3. The sum of the digits of this number that we are required to find has a digit sum of: $0 + 1 + 2 + 3 + 4 + 5 + 6 + 7 + 8 + 9 = 45$, which would indicate that the number is divisible by

3. Therefore, it would be impossible to have a ten-digit number consisting of different digits that is not divisible by 3. Consequently, this can never be a prime number!

SOLUTION FOR PROBLEM 27

The solution to this problem will require some algebraic manipulation particularly using exponents. This will give the reader an appreciation for the power that mathematics shows when we see how algebra can help us get a better understanding of the number of properties.

Since $72 = 8 \cdot 9 = 2^3 \cdot 3^2$, we merely have to show that the number z is divisible by both 8 and 9.

We will begin by considering divisibility by 9. Since the number n is even, we can represent it as $n = 2k$, where the natural number $k > 0$.

Check for divisibility by 9:

Since n is an even number and $k > 0$, we can state the following:

$z = 3^n + 63 = 3^{2k} + 3^2 \cdot 7 = 3^2 \cdot 3^{2k-2} + 3^2 \cdot 7 = 9 \cdot (3^{2k-2} + 7)$.

(Because $k > 0$, it follows that $2k - 2 \geq 0$, to that $3^{2k-2} + 7$ must be a natural number.) This tells us that z must be a multiple of 9.

Check for divisibility by 8:

If we let $2m = 2k - 2 \geq 0$, then $z = 9 \cdot (3^{2k-2} + 7) = 9 \cdot (3^{2m} + 7)$ $= 9 \cdot (3^{2m} - 1 + 8)$.

We can factor $3^{2m} - 1 = (3^m - 1)(3^m + 1)$.

We note that $3^m - 1$ and $3^m + 1$ are two consecutive even numbers, which implies that not only are both divisible by 2, but one of them is also divisible by 4. Therefore, the product $(3^m - 1)(3^m + 1)$ must be divisible by 8. Since z is divisible by the relatively prime numbers 9 and by 8, then z must also be divisible by 72.

SOLUTION FOR PROBLEM 28

A point on the circumference of a circle with diameter d that rotates α degrees will turn $\frac{\alpha}{360} \cdot \pi d$, where the circumference is πd. To determine the motion that we require here of α degrees is $\frac{90}{360} \cdot \pi \cdot 1 = \frac{\alpha}{360} \cdot \pi \cdot 9$, which gives us $\alpha = 10°$.

SOLUTION FOR PROBLEM 29

Most readers would immediately set up the following two equations: $x + y = 2$, and $xy = 3$. Typical algebra training leads us to prepare to solve these two linear equations simultaneously. That would have us change the first equation to read $y = 2 - x$, and then substitute this value for y into the second equation to get: $x(2 - x) = 3$, which leads to the quadratic equation $x^2 - 2x + 3 = 0$. Using the quadratic formula[4] to solve this equation, we find that $x = 1 \pm i\sqrt{2}$. We would then need to find the value of y. Then we would have to take the reciprocals, and then to add them to get the required answer, which is a rather-cumbersome method of solution. The curious part of this original problem is that it can be solved very simply, if we focus on what we have been asked to find, and we do not become distracted by finding the values of x and y.

We have been asked to find the sum of the reciprocals. That is, $\frac{1}{x} + \frac{1}{y}$. So let's find the sum of the reciprocals: $\frac{1}{x} + \frac{1}{y} = \frac{x+y}{xy}$, which essentially gives us the answer, since we know both the numerator and denominator from the given information: $x + y = 2$, and $xy = 3$. Hence, $\frac{1}{x} + \frac{1}{y} = \frac{x+y}{xy} = \frac{2}{3}$, and our problem is solved. Notice, by working backward we achieved a very elegant solution to a problem that otherwise would have been rather complicated to solve.

SOLUTION FOR PROBLEM 30

This may appear to be somewhat complicated, but with a bit of clever manipulation, we can determine that it has no negative roots—without even solving the equation. To do this, we rewrite this equation in the following form: $5x^3 + 7x = x^4 - 4x^2 + 4$, which is then $5x^3 + 7x = (x^2 - 2)^2$. Suppose we consider a negative value for x, then the left side of the equation will be negative, but the right side will always be positive or equal to zero. Therefore, no negative value can satisfy this equation.

SOLUTION FOR PROBLEM 31

Although one may be tempted to use a calculator to evaluate his expression, all too often our expectations for the calculator are overestimated and the result comes back with an "error" message. With the knowledge of powers of three, the problem can be rather cleverly evaluated as follows:

(a) $\dfrac{729^{35} - 81^{52}}{27^{69}} = \dfrac{(3^6)^{35} - (3^4)^{52}}{(3^3)^{69}} = \dfrac{3^{210} - 3^{208}}{3^{207}} = \dfrac{3^{208} \cdot (3^2 - 1)}{3^{207}} = 3 \cdot 8 = 24.$

This expression can be simplified by breaking the numbers up into prime factors as follows:

(b) $\dfrac{6 \cdot 27^{12} + 2 \cdot 81^9}{8000000^2} \cdot \dfrac{80 \cdot 32^3 \cdot 125^4}{9^{19} - 729^6} = \dfrac{2 \cdot 3 \cdot (3^3)^{12} + 2 \cdot (3^4)^9}{(3^2)^{19} - (3^6)^6} \cdot \dfrac{2^4 \cdot 5 \cdot (2^5)^3 \cdot (5^3)^4}{(2^3 \cdot 2^6 \cdot 5^6)^2}$

$= \dfrac{2 \cdot 3^{37} + 2 \cdot 3^{36}}{3^{38} - 3^{36}} \cdot \dfrac{2^{19} \cdot 5^{13}}{2^{18} \cdot 5^{12}} = \dfrac{2 \cdot 3^{36}(3+1)}{3^{36}(3^2 - 1)} \cdot 2 \cdot 5 = \dfrac{2(3+1)}{3^2 - 1} \cdot 2 \cdot 5 = 10.$

SOLUTION FOR PROBLEM 32

The typical solution to this problem could be to simulate the actual tournament by beginning with twelve randomly selected teams playing a second group of twelve teams—with one team drawing a bye. This would then continue with the winning teams playing each other as shown here.

Any **12 teams** versus any other **12 teams**, which leaves **12 winning teams** in the tournament.

6 winners versus **6 other winners**, which leaves **6 winning teams** in tournament.

3 winners versus **3 other winners**, which leaves **3 winning teams** in tournament.

3 winners + **1 team** (which drew a bye) = **4 teams**.

2 remaining teams versus **2 remaining teams**, which leaves **2 winning teams** in the tournament.

1 team versus **1 team** to get a **champion**!

Now counting up the number of games that have been played (figure 3.35) we get:

Teams playing	**Games played**	Winners
24	**12**	12
12	**6**	6
6	**3**	3
3 + 1 bye = 4	**2**	2
2	**1**	1

Figure 3.35

The total number of games played is: $12 + 6 + 3 + 2 + 1 = 24$.

This seems like a perfectly reasonable method of solution, and certainly it's a correct one. However, an alternative, and vastly easier, way to look at this problem is to consider the losers rather than winners. In that case, we ask ourselves, how many losers must there have been in this competition in order to get one champion? Clearly, there had to be twenty-four losers. To get twenty-four losers, there needed to be twenty-four games played. And with that, the problem is solved. Looking at the problem from an alternative point of view is a curious approach that can be useful in a variety of contexts.

SOLUTION FOR PROBLEM 33

From our experience from the previous problem, we know that there needed to be thirty-one games played in order to get a winner of the tournament. The number of possible pairings of the thirty-two players is

$$_{32}C_2 = \frac{32 \cdot 31}{1 \cdot 2} = 16 \cdot 31.$$

Therefore the probability of these two players meeting in the tournament of thirty-one games is $\frac{31}{16 \cdot 31} = \frac{1}{16} = 0.0625$, which is $6\frac{1}{4}$ percent. The complicated made simple!

SOLUTION FOR PROBLEM 34

It is very easy to get bogged down with a problem like this one. Some readers may begin to make a table showing the amount of wine and water in the bottle on each day, and then attempt to compute the proportional amounts of each type of liquid David drinks on any given day. We could more easily resolve the problem by examining it from another point of view, namely, how much water does David add to the mixture each day? Since he eventually empties the bottle (on the sixteenth day), and it held no water to begin with, he must have consumed all the water that was put into the bottle. On the first day, David added one ounce of water. On the second day, he added two ounces of water. On the third day, he added three ounces of water. On the fifteenth day, he added fifteen ounces of water. (Remember, no water was added on the sixteenth day.) Therefore, the number of ounces of water David consumed was

$$1 + 2 + 3 + 4 + 5 + 6 + 7 + 8 + 9 + 10 + 11 + 12 + 13 + 14 + 15 = 120 \text{ ounces.}$$

While this solution is indeed valid, a slightly simpler analogous problem to consider would be to find out how much liquid David drank altogether, and then simply deduct the amount of wine, namely sixteen ounces. Thus,

$$1 + 2 + 3 + 4 + 5 + 6 + 7 + 8 + 9 + 10 + 11 + 12 + 13 + 14 + 15 + 16 - 16 = 120.$$

David consumed 136 ounces of liquid, of which 120 ounces was the water.

SOLUTION FOR PROBLEM 35

The typical response to this question is to simulate the action in the problem by counting systematically:

Member #1 calls three other members; total members contacted = 4.
Members #2, #3, #4 each make three calls; total members contacted
= 4 + 9 = 13.
Members #5–#13 each make three calls; total members contacted
= 4 + 9 + 27 = 40.
Members #14–#33 each make three calls; total members contacted
= 4 + 9 + 27 + 60 = 100.

Since thirty-three members had to make telephone calls to reach all one hundred members, there were sixty-seven members who did not have to make any calls.

Here is the more clever—or elegant—solution: The problem can be solved in a simpler way by working backward. We know that after the first member has been contacted, there are ninety-nine additional members who have to be contacted. This requires thirty-three members each making three telephone calls. This leaves sixty-seven members who need not make any calls.

SOLUTION FOR PROBLEM 36

There are a number of ways to approach this problem. Unfortunately, some of these lead the problem solver astray. A clever approach is one where we use extremes.

Rather than use a tablespoon, we could just as easily use a spoon (or ladle) large enough to contain a half-liter of liquid—in other words the entire amount. Repeating the requirements of this problem, but replacing the tablespoon with the half-liter ladle, we have the following situation. By taking one-half liter of the red wine (which is now the entire quantity of red wine) and pouring it into the white-wine bottle, we find that we have a mixture that is 50 percent red wine and 50 percent white wine. We then take the half-liter ladle, fill it from this bottle, removing a half-liter from this bottle, and pour it back into the bottle that contained all red wine and was emptied. Therefore, there is as much white wine in the red wine as there is red wine in the white wine.

The other extreme to be considered is if the spoon being used is so small that it can only contain zero amount of liquid. This would allow us to conclude that there is as much red wine in the white wine as white wine and red wine—namely, a zero amount.

We can also look at this problem from a logical point of view. We begin with two full bottles, one containing white wine, and the other containing red wine. Whatever amount of red wine we pour into the white-wine bottle is the amount missing from the red wine bottle. When we refill the red wine bottle, that missing amount will be recouped; if we take none of that red wine in the refill, then both bottles have the same amount of wine of the other color or all of the red wine back in the refill, then both bottles have 0 percent of the wine of the other color, or any amount in between. This leads us to the same conclusion that there is as much red wine in the white-wine bottle as white wine in the red-wine bottle.

SOLUTION FOR PROBLEM 37

Without an auxiliary line, this problem would be difficult to solve. We will also add three additional circles to the diagram (figure 3.36) to make our solution more easily reached. We now draw this auxiliary line through the center of the lower left circle and containing the center of the upper right circle. We have thereby partitioned these circles into two equal-area parts. In particular, the five original circles are now partitioned equally. For the meticulous reader, we offer the fact that the angle α that the auxiliary line made with the horizontal is[5]: $\alpha = \arctan \frac{1}{3} \approx 18.43°$.

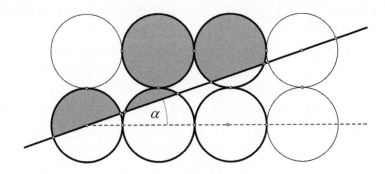

Figure 3.36

SOLUTION FOR PROBLEM 38

This problem can be solved directly with elementary trigonometry; however, it is not a very elegant solution. Below is a quick summary of how the sum of the three angles we seek is 90°.

$\tan \beta = \frac{1}{2} = 0.5$, that is, $\beta = \arctan \frac{1}{2} \approx 0.4636476090$,
   therefore $\beta = 26.56505118 \ldots °$,
$\tan \gamma = \frac{1}{3} = 0.\overline{3}$, that is, $\gamma = \arctan \frac{1}{3} \approx 0.3217505543$,
   therefore $\gamma = 18.43494882 \ldots °$.

Consequently, since $\alpha = 45°$, it follows that because $\beta + \gamma = 45°$, we have $\alpha + \beta + \gamma = 90°$.

This problem lends itself to many rather-elegant solutions. We offer some here, which do not need much explanation. In the first one (see figure 3.37), we construct two squares, *BFPH* and *BFDQ*. We already know that $\alpha = 45°$. We will let $\angle HDP = \delta$, and similarly we notice that $\gamma + \delta = 45°$ ($= \alpha$).

Since $\angle A = \angle P = 90°$ and $\frac{AH}{AC} = \frac{PH}{PD} = \frac{1}{2}$, we have $\triangle ACH \sim \triangle PDH$. Therefore, $\beta = \delta$, this enables us to state that $\gamma + \delta = \gamma + \beta = 45°$, or put another way, we get $\alpha + \beta + \gamma = 90°$, which is what we sought in the first place.

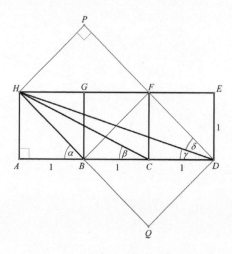

Figure 3.37

Another elegant solution involves other supportive auxiliary lines, as shown in figure 3.38. We notice here that both angles marked β fall between mutually parallel lines and therefore are equal. Also, the two angles marked γ are inscribed angles intercepting the same arc. Therefore, they too are equal. We can then see very clearly that the three angles have a sum $\alpha + \beta + \gamma = 90°$.

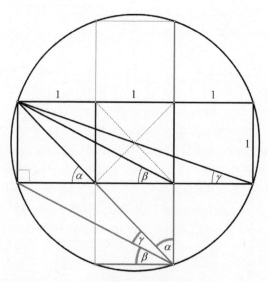

Figure 3.38

Another rather-quick solution requires knowing a trigonometric formula for the tangent of the sum of two angles, namely,

$$\tan(\beta+\gamma)=\frac{\tan\beta+\tan\gamma}{1-\tan\beta\cdot\tan\gamma}=\frac{\frac{1}{2}+\frac{1}{3}}{1-\frac{1}{2}\cdot\frac{1}{3}}=\frac{\frac{5}{6}}{\frac{5}{6}}=1.$$

Therefore, the sum of the angles $\beta+\gamma$ is exactly 45°, which essentially solves the problem since we were not really interested in the individual degree measures.

There are many more elegant solutions to this rather well-known problem, and we leave the reader to discover some of them independently.

SOLUTION FOR PROBLEM 39

This problem appears to be rather complicated, since the result seems counterintuitive. Remember the point P is any point along the diameter, and yet the line segment AB is to be shown to be the same length as the radius. We begin our demonstration by constructing auxiliary lines as shown in figure 3.39. Since parallel lines cut off equal arcs along a circle we can conclude—from the parallel lines drawn in the figure—that the arcs AX, BY, and CZ are equal. Therefore we can conclude that the arcs AB and XY are equal, as are their respective chords: $AB = XY$. However, the chord XY is the third side of an isosceles triangle whose vertex angle $\angle XMY = 60°$. That makes the triangle XMY equilateral. Thus, we have shown that the chord XY is equal to the radius of the circle, since it is equal to XM, and it follows that AB is equal to the radius of the circle as well.

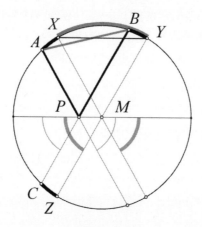

Figure 3.39

SOLUTION FOR PROBLEM 40

The divisors of 360 are 1, 2, 3, 4, 5, 6, 8, 9, 10, 12, 15, 18, 20, 24, 30, 36, 40, 45, 60, 72, 90, 120, 180, and 360.

We could simply get the sum of the reciprocals, namely

$$\frac{1}{1}+\frac{1}{2}+\frac{1}{3}+\frac{1}{4}+\frac{1}{5}+\frac{1}{6}+\frac{1}{8}+\frac{1}{9}+\frac{1}{10}+\frac{1}{12}+\frac{1}{15}+\cdots+\frac{1}{120}+\frac{1}{180}+\frac{1}{360}$$

to obtain the required answer. However, this would be a daunting task.

Sometimes a clever method of solution is to look for a pattern, or to solve a similar analogous problem with more-manageable numbers. One way of looking for a pattern would be to consider a smaller number than 360 and work with that to see if a pattern may emerge.

Suppose we repeat the same problem with the number 12. We find that the sum of the divisors of 12 is $1 + 2 + 3 + 4 + 6 + 12 = 28$. The sum of the reciprocals of the divisors of 12 is $\frac{1}{1}+\frac{1}{2}+\frac{1}{3}+\frac{1}{4}+\frac{1}{6}+\frac{1}{12}=\frac{12}{12}+\frac{6}{12}+\frac{4}{12}+\frac{3}{12}+\frac{2}{12}+\frac{1}{12}=\frac{28}{12}$. There could be a pattern here that we can justify with other such numbers. That is, the sum of the reciprocals appears to be a fraction whose numerator is the sum of the divisors divided by the original number. Naturally, we should make sure that this pattern actually holds for all numbers. Once we can justify this as a correct pattern, we can apply it to the given problem.

We simply take the sum of the divisors, which is 1,170, and divide it by the number itself, 360. Thus, the answer to the problem is $\frac{1170}{360} = \frac{13}{4} = 3.25$. This shows how determining a pattern can make a seemingly complicated situation rather simple.

SOLUTION FOR PROBLEM 41

(a) Some people may begin to attack this problem by entering the powers of 8 into their calculators. However, they should soon realize that most calculators would not permit them to arrive at an answer of that magnitude, since the display will run out of digit space before they reach their goal. Thus, we must look for another approach. Let's examine the increasing powers of 8 and see if there is a pattern in the last digit, which may be of use.

$8^1 =$ **8**	$8^5=$ 32,76**8**	$8^9 =$ 134,217,72**8**	
$8^2 =$ 6**4**	$8^6=$ 262,14**4**	$8^{10} =$ 1,073,741,82**4**	
$8^3 =$ 51**2**	$8^7 =$ 2,097,15**2**	$8^{11} =$ 8,589,934,59**2**	
$8^4 =$ 4,09**6**	$8^8 =$ 16,777,21**6**	$8^{12} =$ 68,719,476,73**6**	

Notice the pattern that emerges—the units digits repeat in cycles of four powers. It would appear that we can apply this pattern to our original problem. Since the exponent we are interested in is 19, which gives a remainder of 3 when divided by 4. Thus, the terminal digit of 8^{19} should be the same as those of 8^{15}, 8^{11}, 8^7, 8^3, which we recognize as 2.

By the way, for the skeptical reader, here is the actual value of 8^{19} = 144,115,188,075,855,87**2**.

(b) Let's examine the increasing powers of 7 and see if there is a pattern that may be of use.

$7^1 =$ **7**	$7^5 =$ 16,80**7**	$7^9 =$ 40,353,60**7**	
$7^2 =$ 4**9**	$7^6 =$ 117,64**9**	$7^{10} =$ 282,475,24**9**	
$7^3 =$ 34**3**	$7^7 =$ 823,54**3**	$7^{11} =$ 1,977,326,74**3**	
$7^4 =$ 2,40**1**	$7^8 =$ 5,764,80**1**	$7^{12} =$ 13,841,287,20**1**	

Following this pattern, we have: the exponent 197 leaves a remainder of 19 (by division through 4) and so the units digit of 7^{197} should be the same as 7^1, which is 7. We can check this answer, if we have the time, and we will find that $7^{197} =$ 305009862720535194606965003259965412822716867351901 855975222742974785007796625721626075294895316736160147 6748 76167531025482891555209434145427135692925359082642491 43207.

SOLUTION FOR PROBLEM 42

Again, some readers will attempt to solve this problem by using their calculators. This is a formidable task, and an error can often be expected! The beauty in this problem is not simply getting the answer but, rather, finding the *path* to the answer. Once again, let's utilize the strategy that requires us to look for a pattern. We must examine the patterns that exist in the powers of three different sets of numbers. Practice in doing this will help familiarize you with the cyclical pattern for the final digits of the powers of numbers.

For powers of 13, we obtain:

$13^1 =$	1**3**	$13^5 =$	371,29**3**
$13^2 =$	16**9**	$13^6 =$	4,826,80**9**
$13^3 =$	2,19**7**	$13^7 =$	62,748,51**7**
$13^4 =$	28,56**1**	$13^8 =$	815,730,72**1**

The units digits for powers of 13 repeat as 3, 9, 7, 1, 3, 9, 7, 1, . . . in cycles of four. Thus, 13^{25} has the same units digit as 13^1 or 13, namely 3.

For powers of 4, we obtain:

$4^1 =$	**4**	$4^5 =$	1,02**4**
$4^2 =$	1**6**	$4^6 =$	4,09**6**
$4^3 =$	6**4**	$4^7 =$	16,38**4**
$4^4 =$	25**6**	$4^8 =$	65,53**6**

The units digits for powers of 4 repeat as 4, 6, 4, 6, 4, 6, . . . in cycles of two. Thus, 4^{81} has the same units digit as 4^1, which is 4.

The units digit for powers of 5 must be the numeral 5 (i.e., $\underline{5}$, 2$\underline{5}$, 12$\underline{5}$, 62$\underline{5}$, etc.).

The sum we are looking for is $3 + 4 + 5 = 12$, which has a units digit of 2.

$(13^{25} + 4^{81} + 5^{411} = $ 18909140209225186878994290201593514880713960 89867573664788946748703328294969573225030606559705573533646512 471927516829853208416210445483552501131860670581294923064484 99 53763685246250187369017353959030115461205773438385108215717621 32234502563535580184937538282843952167480451790116847397 2$\underline{2}$)

You might examine the cyclical nature of the units digits when raising any number to powers. Are the powers of all single-digit numbers cyclical? What are the cycles?

SOLUTION FOR PROBLEM 43

This problem cannot be done on the calculator, since the answer will contain more places than the display permits. It can be done manually, although the computation often leads to an error due to the large number of zeros in the answer. We might, however, examine the answers we obtain by starting with a small divisor, increasing the divisor, and seeing if a usable pattern emerges. (See figure 3.40.)

	Number of Zeros after the 5	Quotient	Number of Zeros after the Decimal and before the 2
1÷5	0	.2	0
1÷50	1	.02	1
1÷500	2	.002	2
1÷5000	3	.0002	3
. . .			
1÷500000000000	11	.000000000002	11

Figure 3.40

The correct answer is now easily found. The number of zeros after the decimal point and before the 2 is the same as the number of zeros in the divisor,

$$\frac{1}{500,000,000,000} = 2 \cdot 10^{-12} = 0.2 \cdot 10^{-11} = 2 \cdot 10^{-12}.$$

SOLUTION FOR PROBLEM 44

One could actually cube all the integers from 1 through 10, and then take the sum. If carefully done with the aid of a calculator, this should yield the correct answer. However, if we do not have a calculator handy, the multiplication and addition could prove quite cumbersome and messy! Let's see how we might solve the problem by searching for a pattern. We will organize our data with a table:

1^3	$= (1)$	$= 1$	$= 1^2$
$1^3 + 2^3$	$= (1 + 8)$	$= 9$	$= 3^2$
$1^3 + 2^3 + 3^3$	$= (1 + 8 + 27)$	$= 36$	$= 6^2$
$1^3 + 2^3 + 3^3 + 4^3$	$= (1 + 8 + 27 + 64)$	$= 100$	$= 10^2$

Notice that the number bases in the final column (namely $1, 3, 6, 10, \ldots$) are the *triangular numbers*. The nth triangular number (see chapter 1, notes 8 and 23) is formed by taking the sum of the first n integers. That is, the first triangular number is 1. The second triangular number, $3 = (1 + 2)$.

The third triangular number, $6 = (1 + 2 + 3)$. The fourth triangular number, $10 = (1 + 2 + 3 + 4)$, and so on.

Thus, we can rewrite our problem as follows:

1^3	$= (1)^2$	$= 1^2$	$= 1$
$1^3 + 2^3$	$= (1 + 2)^2$	$= 3^2$	$= 9$
$1^3 + 2^3 + 3^3$	$= (1 + 2 + 3)^2$	$= 6^2$	$= 36$
$1^3 + 2^3 + 3^3 + \ldots + 9^3 + 10^3$	$= (1 + 2 + 3 + \ldots + 9 + 10)^2$	$= 55^2$	$= 3025$

By this point you should have gotten a "feel" for the advantage of looking for a pattern in solving a problem. It may take some effort to find a pattern, but when one is discovered, it not only simplifies the problem greatly but also once again demonstrates the beauty of mathematics.

SOLUTION FOR PROBLEM 45

Mostly through trial and error, some may stumble on the right answer, namely, the North Pole. To test this answer, try starting from the North Pole and travel south one mile, and then west one mile; that takes you along a latitudinal line, which remains equidistant from the North Pole (i.e., one mile from it). Then, traveling one mile north gets you back to where you began, the North Pole. (See figure 3.41.)

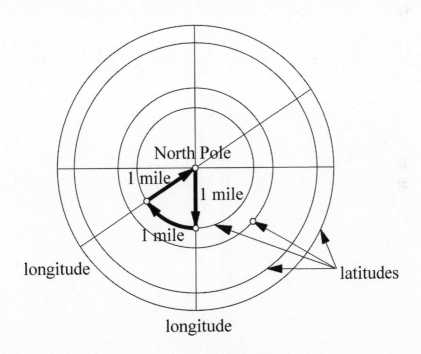

Figure 3.41

Most people familiar with this problem feel a sense of completion. Yet we can ask: Are there other such starting points where we can take the same three "walks" and end up at the starting point? The answer, surprising enough, is yes.

One set of starting points is found by locating the latitudinal circle, which has a circumference of one mile, and is nearest the South Pole. From this latitudinal circle, walk one mile north (naturally, along a great circle route, i.e., a circle whose center is at the center of the earth—a sphere), and form another latitudinal circle. Any point along this second latitudinal circle will qualify. Let's try it.

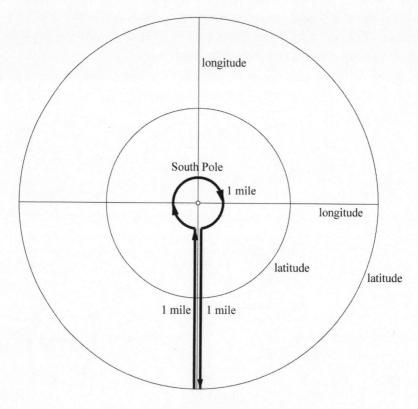

Figure 3.42

Begin on this second latitudinal circle (the one farther north). Walk one mile south (which takes you to the first latitudinal circle), then walk one mile west. This takes you exactly once around the circle. Then, by walking one mile north, you will be back to the starting point. (See figure 3.42.)

Suppose the second latitudinal circle, the one we would walk along, would have a circumference of $\frac{1}{2}$ mile. We could still satisfy the given instructions. Yet this time we would walk around the circle *twice* to get back to our original starting point. If the second latitudinal circle had a circumference of $\frac{1}{4}$ mile, then we would merely have to walk around this circle *four* times to get back to the starting point of this circle, and then go north one mile to the original starting point.

At this point we can take a giant leap to a generalization that will lead us to many more points that satisfy the original stipulations. Actually, an infinite number of points! This set of points can be located by beginning with the latitudinal circle, located nearest the South Pole, and has an $\frac{1}{n}$th-mile circumference, so that an *n*-mile walk east will take you back to the point on the circle at which you began your walk on this latitudinal circle. The rest is the same as before, that is, walking one mile south, and then later one mile north. Is this possible with latitude circle routes near the North Pole? Yes, of course!

SOLUTION FOR PROBLEM 46

Most people would conjecture that by staying on that course, the plane would end up at the point at which it started—assuming it had enough fuel to make the journey. This assumption is incorrect!

From a logical point of view, if the plane is to return to its original position, it cannot continuously fly in a northeastern direction, since it would probably have to be heading southwest—no longer northeast. The path the plane would take is called a *loxodrome*.[6] (See figure 3.43.)

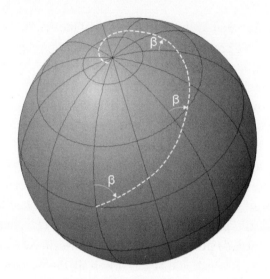

Figure 3.43

The path is actually a never-ending spiral approaching the North Pole, but actually (theoretically) never reaching it. As you can see in figure 3.43, the path gets ever so much closer to the North Pole, yet we cannot show how it just spins around that pole without reaching it.

The spiral is generated by the course determined by the angle β between the latitude and the longitude line, which remains constant throughout.

Now that we have shown a counterintuitive answer to the original question of the plane's path, we might just take a closer look at the loxodrome spiral whose equation could be shown as

$\lambda = \ln\left[\tan\left(\frac{\pi}{4}+\frac{\varphi}{2}\right)\right]$, where λ and φ represent the latitude and longitude measures.

For $\varphi = 0°$, we get $\lambda = 0°$. However, as φ approaches 90°, then λ approaches infinity. That means that the loxodrome will spiral around the North Pole indefinitely. Despite that, the length of the spiral is finite and is equal to $1 = \frac{\sqrt{2}}{2}\pi \cdot r \approx 14,153$ km, with the radius of the earth assumed to be $r = 6,371$ km.

We should also note that on a Mercator[7] projection map, a loxodrome is a straight line, which can be drawn on such a map between any two points on earth without going off the edge of the map. But, theoretically, a loxodrome can extend beyond the right edge of the map, where it then continues at the left edge with the same slope (assuming that the map covers exactly 360 degrees of longitude).

SOLUTION FOR PROBLEM 47

The curiosity in this problem is that the solution lies in the orientation of the figures. In order to find the relative sizes of the two equilateral triangles, we simply rotate the inside triangle to the position shown in figure 3.44. This gives us an immediate answer to our question. We notice that we now have four congruent equilateral triangles comprising the large equilateral triangle. Therefore, the ratio of the areas of the inscribed equilateral triangle to the circumscribed equilateral triangle is 1:4, and the ratio of the sides is then 1:2.

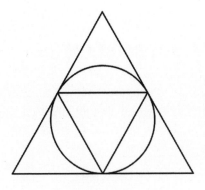

Figure 3.44

SOLUTION FOR PROBLEM 48

The solution to this problem becomes obvious and is made trivial by simply drawing the diagonals of the rectangle. (See figure 3.45.) If we consider as the hypotenuse any one of the right triangles formed by one of these diagonals, we find that a side of the rhombus, which is a line segment joining the midpoints of two sides of that right triangle, is half the length of the hypotenuse (i.e., the diameter of the circle) and, therefore, has length 6. As is sometimes the case when a mathematical challenge is presented, there may be extraneous information provided. This is the case here, where the ratio of the sides of the rectangle is irrelevant to the solution of this problem.

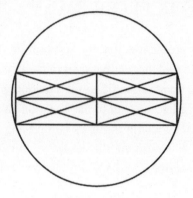

Figure 3.45

SOLUTION FOR PROBLEM 49

As the problem is presented in figure 3.46, we can approach the problem by drawing some additional lines and applying the Pythagorean theorem. This will bring us the necessary dimensions to allow us to get the relationship between the two areas.

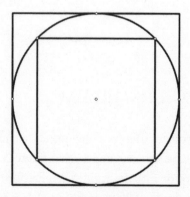

Figure 3.46

However, there is a much-simpler way to approach this problem, and that is by noticing that there is a definite symmetry inherent in the diagram. And using that to our advantage makes the problem trivial. By rotating the inscribed square 45°, the configuration will appear as in figure 3.47. Here we notice that each quarter of the smaller square contains half the area of a quarter of the larger square. Therefore, the smaller square is one half the area of the larger square, or, put another way, the ratio of the areas is 1:2. It is curious how considering symmetry can be very useful for solving an apparently difficult problem in a rather-simple fashion.

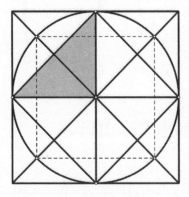

Figure 3.47

SOLUTION FOR PROBLEM 50

Once again we had the situation where the key is to look beyond the perhaps-distracting information. Once we realize that the rectangle has equal diagonals, and we know that the diagonal AC is also a radius of the quarter circle, namely, 10, then it follows that $DB = 10$ as well. (See figure 3.48.)

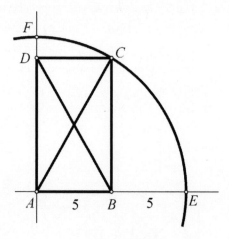

Figure 3.48

SOLUTION FOR PROBLEM 51

When asked to use four straight lines to connect the nine dots in the figure below, and doing so without lifting the pencil when drawing the lines, most people would go along the four sides of the square only to realize that the center dot has been omitted. The next step would be to include the center dot and then find that other dots have been omitted. For most people there appears to be a psychological barrier to have the lines *extend* beyond the square. However as you can see from figure 3.49, there are two variations of how this can be accomplished. It is curious that most people won't "think outside the box"!

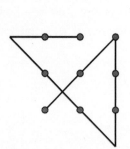 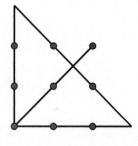

Figure 3.49

SOLUTION FOR PROBLEM 52

It is well known that many people become frustrated very quickly as they attempt to solve this problem. It seems clear that they would have to remove one stick from each (outside) row and each (outside) column, and that is typically done, which leaves the configuration shown in figure 3.50.

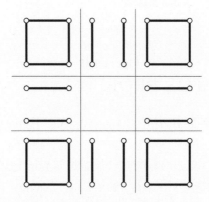

Figure 3.50

However, in figure 3.50 there remain only ten sticks in each outside row and outside column. The question is, how can we have eleven sticks in each outside row and outside column after we have removed one stick from each row and column? Once again, we need to think "outside the box."

We simply take another stick from each outside row and column, as shown in figure 3.51, and place one at each of the corners, and the problem is solved. This may appear trivial, but experience shows that most people do not "stumble" easily on this curious solution.

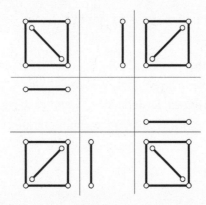

Figure 3.51

SOLUTION FOR PROBLEM 53

The solution, surprisingly, is quite simple. Just take the middle full glass and pick it up.

Then pour the contents of the glass that is now being held, and fill the right-most empty glass with this water. Place this now-empty glass back to its original position. That's it! The arrangement now is such that every empty glass is next to a full glass, and every full glass is next to an empty glass.

SOLUTION FOR PROBLEM 54

Typically, a first attempt at a solution involves opening the end link of one chain, joining it to the second chain to form a six-link chain, then opening and closing a link in the third chain and joining it to the six-link chain to form a nine-link chain. By opening and closing a link in the fourth chain and joining it to the nine-link chain, a twelve-link chain, which is *not* a circle, is obtained. Thus, this typical attempt usually ends unsuccessfully. Most attempts usually involve other combinations of opening/closing one link of each of various chain pieces to try to join them together to get the desired result, but this approach will not work.

Let's look at this from another point of view. Instead of continually trying to open and close *one* link of each chain piece, a different point of view would involve opening *all* the links in one chain and using those links to connect the remaining three chain pieces together into the required circle chain.

This quickly gives the successful solution.

SOLUTION FOR PROBLEM 55

The normal reaction is to try to find the area of each of the parallelograms, and then try to make the comparison. This will not lead to a simple solution! Curiously, this problem is solved very easily by drawing the line segment *BG*. We now see in figure 3.52 that the two parallelograms have the dark shaded triangle *ABG* as part of their respective areas.

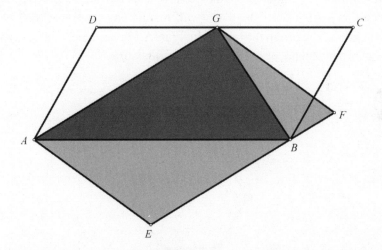

Figure 3.52

Let us first focus on triangle *ABG* as it relates to parallelogram *ABCD*. The triangle and the parallelogram share the same base, *AB*, and the same altitude, from point *G* to *AB*.

Therefore, $Area \triangle ABG = \frac{1}{2} Area ABCD$.

A similar argument can be made for the relationship between triangle *ABG* and parallelogram *AEFG*, since the two share the same base, *AG*, and the same altitude from point *B* to side *AG*. Hence, once again *Area△ABG* $= \frac{1}{2} AreaAEFG$. Consequently, since each of the two parallelograms has an area twice that of triangle *ABG*, the two parallelograms are equal in area.

As the problem was presented in general terms, with the points *G* and *B* located on lines *CD* and *EF*, respectively, and not specified as to where on the lines the points were placed, we could have actually placed them on points *D* and *F*, respectively, and the problem would have become trivial! In other words, the area of parallelogram *AEFG* is independent upon where the point *G* is located on line segment *CD*.

SOLUTION FOR PROBLEM 56

To minimize the number of points that have to be moved, we need to search within this pattern for a shape that will remain constant regardless of the orientation of the original triangle. In figure 3.53, we notice the hexagon is symmetric independent of the triangle's orientation.

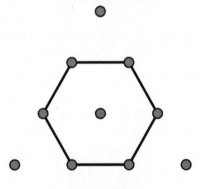

Figure 3.53

Therefore, we need to move the three dots that are not part of the hexagon (i.e., those at the vertices of the equilateral triangle), as shown in figures 3.54a and 3.54b.

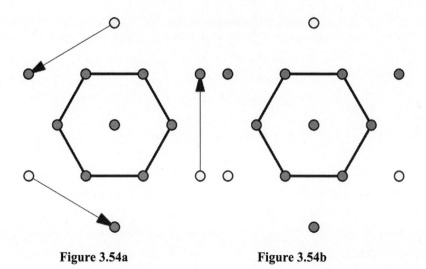

Figure 3.54a   **Figure 3.54b**

SOLUTION FOR PROBLEM 57

The typical response to this question is one hundred days, reflecting his progress of one foot per day. However, after ninety-seven days, the monkey has progressed ninety-seven feet. Therefore, on the morning of the ninety-eighth day, he will have reached the top by climbing the remaining three feet needed. Thus, rather than the one hundred days that one expects would be required, here only ninety-seven and a half days were required.

SOLUTION FOR PROBLEM 58

Since $\frac{3}{2}$ hens work for $\frac{3}{2}$ days, we can create a unit of work called "hen-days," which would be the product of the number of hens laying eggs for certain number of days. In this case, we would have $\frac{3}{2} \cdot \frac{3}{2} = \frac{9}{4}$ hen-days. The second job, where six hens lay eggs for eight days would then comprise forty-eight hen-days. Letting x equal the number of eggs being laid in forty-eight hen-days, we can set up the following proportion:

$$\frac{\frac{9}{4}\,\text{hen-days}}{48\,\text{hen-days}} = \frac{\frac{3}{2}\,\text{eggs}}{x}$$

Then $\frac{9}{4}x = 48 \cdot \frac{3}{2} = 72$, and $x = 32$ eggs.

SOLUTION FOR PROBLEM 59

Most people would tend to approach this problem by simply guessing at the answer, and keep "pouring" back and forth in an attempt to arrive at the correct answer, a sort of "unintelligent" guessing-and-testing routine. However, the problem can be solved in a more organized manner by using a strategy of working backward. We need to end up with seven liters and (obviously) it would have to be in the eleven-liter can. When we do this, we would be leaving a total of four empty liters in the eleven-liter can (see figure 3.55). But how do we arrange for four empty liters in that can?

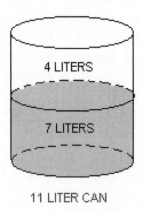

4 LITERS

7 LITERS

11 LITER CAN

Figure 3.55

To obtain four liters, we must leave one liter in the five-liter can. Now, how can we obtain one liter in the five-liter can? Fill the eleven-liter can and pour from it by twice filling the five-liter can and discarding these two amounts of five liters of water. This leaves one liter in the eleven-liter can. (See figure 3.56.)

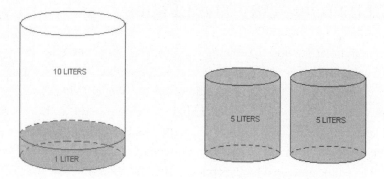

Figure 3.56

Then pour the one-liter into the five-liter can. We are now ready to fill the eleven-liter can and pour off the four liters needed to fill the remainder of the five-liter can. This leaves the required seven liters in the eleven-liter can. (See figure 3.57.)

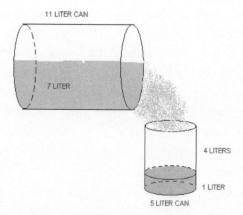

Figure 3.57

Note that problems of this sort do not always have a solution. That is, if you wish to construct additional problems of this sort, you must bear in mind that a solution only exists when the difference of multiples of the capacities of the two given cans can be made equal to the desired quantity.

In this problem, $2 \cdot 11 - 3 \cdot 5 = 7$.

SOLUTION FOR PROBLEM 60

The traditional method for solving this problem, a common one in algebra textbooks, is as follows. The water carries Simon one mile downstream as well as carrying the cork one mile downstream. We let t equal the time the cork travels downstream after the initial ten minutes, and we have r equal the speed of rowing in still water, and we have s equal the rate of the stream. Recalling that the product of rate and time equals the distance traveled, we get the distance traveled by the cork as $(10 + t)s = 1$ mile. Now the distance Simon, who has turned about and is following the cork, must travel is $t(r + s)$, which equals the return distance of his upstream trip, $10(r - s)$, plus the 1 mile that the cork traveled downstream.

Therefore, $t = 10$. So the time the cork had been traveling downstream is $10 + 10 = 20$ minutes, and twenty minutes to travel one mile translates to the stream rate of three miles per hour.

Rather than approach this problem through the traditional methods as we did above, which is common in an algebra course, consider the following. The problem can be made significantly easier by considering the notion of relativity. It does not matter if the stream is moving and carrying Simon downstream or if it is still. We are concerned only with the separation and coming together of Simon and the cork. If the stream were stationary, Simon would require as much time rowing to the cork as he did rowing away from the cork. That is, he would require $10 + 10 = 20$ minutes. Since the cork travels one mile during these twenty minutes, the stream's rate of speed is three miles per hour.

This is a concept (relativity) worth understanding, for it has many useful applications in everyday-life thinking processes. This is, after all, one of the purposes for knowing mathematics.

SOLUTION FOR PROBLEM 61

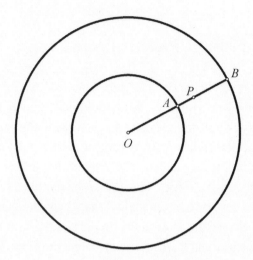

Figure 3.58

The typical (and correct) answer is $\frac{1}{4}$, since we know that the area of the smaller circle is $\frac{1}{4}$ the area of the larger circle. Therefore, if a point is selected at random in the larger circle, the probability that it would be in the smaller circle as well is $\frac{1}{4}$.

However, we could look at this question differently. The randomly selected point P must lie on some radius of the larger circle, say radius *OAB*, where *A* is its midpoint. (See figure 3.58.) The probability that a point P on *OAB* is on *OA* (i.e., in the smaller circle) is $\frac{1}{2}$. Now if we do this for any other point in the larger circle, we find the probability of the point being in the smaller circle as well is $\frac{1}{2}$. This, of course, is not correct. Where was the error made in the second calculation?

The "error" lies in the initial definition of each of two different sample spaces. In the first case, the sample space is the entire area of the larger circle, whereas in the second case, the sample space is the set of points on *OAB*. Clearly, when a point is selected on *OAB*, the probability that the point will be on *OA* is $\frac{1}{2}$. These are two entirely different problems even though (to dramatize the issue) they appear to be the same. Conditional probability is an important concept to understand, and what better way to instill this idea than through a demonstration that shows obvious absurdities?

SOLUTION FOR PROBLEM 62

At first glance, it would appear that there is not enough information given to solve this problem. However, with additional support lines as shown in figure 3.59, several opportunities become available to us. We know that a radius is perpendicular to a tangent at the point of contact (T). Furthermore, a radius perpendicular to a chord divides the chord into two equal segments. Thus, $AT = BT = 4$. We know that the area of the region between the two circles (the "doughnut" shape) can be found by obtaining the difference between the areas of the two circles. We will let the radius of the smaller circle be r and the radius of the larger circle be R. Thus, the area of this region between the two circles equals $\pi R^2 - \pi r^2$ or $Area = \pi (R^2 - r^2)$.

Now, since $OC = R$ and $OT = r$, $CT = R - r$ and $DT = R + r$.

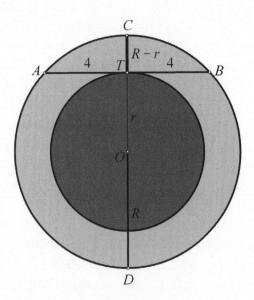

Figure 3.59

Since we know that the products of the segments of two intersecting chords in a circle are equal, we obtain $CT \cdot DT = AT \cdot BT$, or

$$(R - r)(R + r) = 4 \cdot 4$$
$$R^2 - r^2 = 16$$

Therefore, the area A of the region between the two circles is $A = 16\pi$ square units. The problem is solved.

Another way to approach this problem is by drawing line segment OA. We now create a right triangle ATO, in which $R^2 = OA^2 = OT^2 + AT^2 = r^2 + 4^2$, or $R^2 - r^2 = 16$.

Up until now, the methods of solution were rather straightforward and traditional, using the normal tools from elementary geometry. However, to demonstrate the power of mathematics, we can also look at this problem in a more unusual way, that is, by considering an extreme case. Let's assume that the smaller circle gets smaller and smaller until it becomes extremely small—so small that we would see it as a point that coincides with point O. Then AB becomes a diameter of the larger circle, and the area of the region between the two circles becomes the area of the larger circle, which equals $\pi R^2 = 16\pi$. This elegant solution allows this problem to once again demonstrate the value of considering an extreme case.

SOLUTION FOR PROBLEM 63

To determine the ages of the three daughters, the man makes a chart of the possible ages (figure 3.60), that is, the three numbers whose product is 36:

Age of Daughter *a*	Age of Daughter *b*	Age of Daughter *c*	Age Sum *a* + *b* + *c*
1	1	36	38
1	2	18	21
1	3	12	16
1	4	9	14
1	6	6	13
2	2	9	13
2	3	6	11
3	3	4	10

Figure 3.60

From the chart in figure 3.60, we see that with all the possible products of 36, there is only one case where the man would need additional information to figure out the answer; that is, if the sum of their ages is 13. So, when the woman tells the man that she left out an essential piece of information, it must have been to differentiate between the two sums of 13. When she mentioned that her oldest daughter is named Miriam, that indicated that there was only one older daughter, thus eliminating the possibility of twins of age 6. Hence, the ages of the three daughters are 2, 2, and 9.

Here, arithmetic calculation alone did not help us answer the question.

SOLUTION FOR PROBLEM 64

To answer this problem, we need to reach back to our knowledge of trigonometry. That is, we need to recall that the cosine function of a right angle is 0, the cosine function of an obtuse angle is negative, and the cosine of an acute angle is positive.

Therefore, in a right triangle, one of the cosine values will be 0, thereby making the product $\cos\alpha \cdot \cos\beta \cdot \cos\gamma = 0$. Furthermore, in an obtuse triangle, one of the angles will be greater than 90°. Say, for example, $\alpha > 90°$, then $\beta < 90°$ and $\gamma < 90°$, but the product $\cos\alpha \cdot \cos\beta \cdot \cos\gamma < 0$. Finally, for an acute triangle, where all the values of the cosine functions are positive so that cos α > 0 and cos β > 0 and cos γ > 0, then the product cos $\alpha \cdot$ cos $\beta \cdot$ cos γ > 0. Therefore, the triangle described in the original question is an acute triangle.

SOLUTION FOR PROBLEM 65

Rather than to try to construct this entire triangle until you get to the appropriate row, we should look for a pattern from the information given. We notice that the last member of each row is a perfect square. As a matter of fact, it is the square of the row number. We also notice that the first member of each row is also related to the row number as follows:

The nth row has as the last number n^2, and as the first member of this row we have $(n - 1)^2 + 1$.

To find the middle member of this row, we take the average of the two end members:

$$\frac{[(n-1)^2+1]+n^2}{2}=\frac{n^2-2n+1+1+n^2}{2} = n^2 - n + 1.$$

Since the number 2000 is 44^2 < 2000 < 45^2, it must be in the forty-fifth row, where the first member is 1937 and the last member is 2025. The middle member is 1981, which allows you to determine the location of the number 2000 in this row.

SOLUTION FOR PROBLEM 66

We begin by letting t equal the time required for the two runners to meet at noon. Let a represent the speed of the first runner, who starts from town A and who reaches town B at 4 p.m. Let b represent the speed of the second runner, who runs from town B and reaches town A at 9 p.m.

The first runner requires t hours to reach the point at which the two runners meet at noon. The second runner, going the opposite direction, required nine hours to cover the same distance. We can represent this distance—from town

A to the meeting point—in two ways that are equal: $ta = 9b$. In an analogous fashion, we can represent the distance from town B to the meeting point as $tb = 4a$. Now, with a little algebraic foresight, we can multiply these two equations to get $t^2ab = 36ab$. This simplifies to $t^2 = 36$, which gives us $t = 6$. This tells us that sunrise was at 6 a.m.

SOLUTION FOR PROBLEM 67

Here is another problem where the elegance comes out in the alternative solution we present after you have seen the traditional solution. This order of presentation allows you to even better appreciate the elegance of the second solution.

The expected solution would have us drawing BN, and then recognizing that triangle ABN has half the area of triangle ABC (since the median of a triangle partitions it into two equal areas).

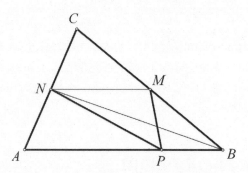

Figure 3.61

Also, since a line joining the midpoints of two sides of a triangle is parallel to the third side, MN is parallel to AB. Therefore, triangle BMN has the same area as triangle PMN because they share the same base MN and have equal altitudes to that base, since their vertices lie on a line parallel to their common base. (See figure 3.61.)

The $Area\ \Delta BMN + Area\ \Delta CNM = Area\ \Delta BCN = \frac{1}{2} Area\ \Delta ABC$.

By substitution, we have *Area* $\triangle PMN$ + *Area* $\triangle CNM$ = $\frac{1}{2}$ *Area* $\triangle ABC$.

That is to say that the area of quadrilateral *MCNP* is one-half of the area of triangle *ABC*.

We could make this problem (practically) trivial by choosing a convenient point, *P*, on *AB*, since no specific point location for *P* on *AB* was given. By considering the extremes, suppose *P* was at one extreme end of *AB*, say at point *B*. In that case, the quadrilateral *MCNP* reduces to become triangle *BCN*, since the length of *BP* is 0. As we mentioned above, triangle *BCN* results from a median partitioning a triangle; hence, the area of triangle *BCN* is one-half the area of triangle *ABC*.

SOLUTION FOR PROBLEM 68

The traditional solution to this problem requires that we recognize the figure in the middle as a regular tetrahedron on equilateral triangle faces. While the cube's edge length is *a*, the edge of the tetrahedron is $b = a\sqrt{2}$ (using the Pythagorean theorem). Further applying the Pythagorean theorem, we find the height *h* of the tetrahedron meets the opposite face at the center of the equilateral triangle. If the length *m* of the median of the base (which is also the height of the equilateral triangle) is $\frac{a\sqrt{6}}{2}$, then $\frac{2}{3}m = \frac{a\sqrt{6}}{3}$. We have a right triangle where this length is that of one leg, and *b* is the hypotenuse of that right triangle. We then have the height of the tetrahedron as $h = \frac{2\sqrt{3}}{3}a$. This then allows us to calculate the volume of the tetrahedron as follows:

$$V = \frac{G \cdot h}{3} = \frac{1}{3} \cdot \frac{b \cdot m}{2} \cdot h = \frac{1}{3} \cdot \frac{a\sqrt{2} \cdot \frac{a\sqrt{6}}{2}}{2} \cdot \frac{2\sqrt{3}}{3}a = \frac{a^3}{3}.$$

However, a much more elegant solution can be arrived at by recognizing that we have four congruent pyramids at each of the corners of the cube. If we subtract the volumes of these four pyramids from the volume of the cube, we will be left with the required volume of the tetrahedron. The base of each of these four pyramids is an isosceles right triangle (area *G*) whose leg is length *a*, and whose hypotenuse is length *b*. The height of the pyramid is also length *a*. Therefore, the volume of the four pyramids is

$$V_{4P} = 4 \cdot \frac{G \cdot h}{3} = \frac{4}{3} \cdot \frac{a \cdot a}{2} \cdot a = \frac{2}{3} a^3.$$

When we subtract this from the volume of the cube we get

$$V = V_{cube} - V_{4P} = \frac{1}{3} a^3.$$

SOLUTION FOR PROBLEM 69

The typical, mistaken, answer is no. Yet you will see that there actually is such a solid figure.

To find a solid figure that has the shape of a square when viewed from three different directions is obviously easy to imagine. It is a cube with edge length 1, which will appear as a square when viewed from the common three directions. A sphere with a diameter of 1 will appear as the circle shown in figure 3.62, when viewed from various directions. A cylinder with a height of length 1 and the base circle with diameter of length 1 can also be viewed from various directions to be seen as a square or a circle as in figure 3.62. Yet, to get one solid that will render these three shapes when viewed from different directions seems nearly impossible and, therefore, usually renders a mistaken answer, as we indicated earlier.

Suppose we now cut these three shapes out of the board, as shown in figure 3.62. Let's see if we can find solid shapes that can be pushed through these figures with a tight fit.

Clearly, a cube can fit through the square as long as the edge length is appropriately sized. A sphere of diameter 1, or a cylinder of diameter 1, can easily be pushed to the circular cutout. A prism with a base congruent to the cutout of the isosceles triangle will also fit through the triangular cutout.

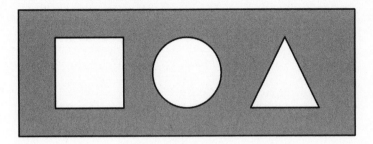

Figure 3.62

However, the solid that we seek must be one that could be pushed through each of these three shapes with the expected tight fit. One possible solid is constructed by taking a cube with a unit length edge, where we then cut a cylinder out of it as shown in figure 3.63. We than remove all the matter that after two planer cuts (cuts that leave planes, as shown in the figure) is not included in a right prism.

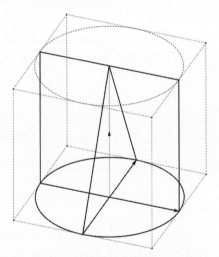

Figure 3.63

In figure 3.64, we see various cut lines on the surface of this figure.

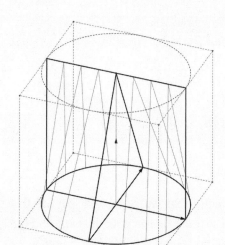

Figure 3.64

Figure 3.65 shows photographs of this solid.

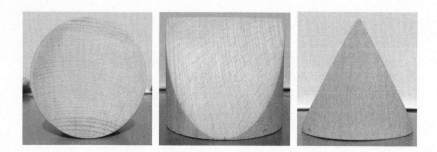

Figure 3.65

SOLUTION FOR PROBLEM 70

We usually begin to think about the likelihood of two people having the same birth date out of a selection of 365 days (assuming no leap year). Is it perhaps two out of 365, or about $\frac{1}{2}$ percent? To begin we can look at the randomly selected group of the first thirty-five presidents of the United States. You may be astonished to learn that there are two presidents with the same birth date: James K. Polk (November 2, 1795) and Warren G. Harding (November 2, 1865). For a group of thirty-five randomly selected people, the probability that two members will have the same birth date is greater than 0.8. For groups of thirty, the probability that there will be a match of birth dates is greater than 0.7, or, put another way, more frequently than in seven of ten times, there ought to be a match of birth dates. What causes this incredible—and counterintuitive—result? Let us investigate this surprising situation.

We begin by realizing that the probability that one selected member of the group matches his own birth date is clearly certainty, or 1. This can be written as $\frac{365}{365}$.

The probability that another member of the group does *not* match the first members birthday is $\frac{365-1}{365} = \frac{364}{365}$.

The probability that a third member of the group does *not* match the first and second members' birthdays is $\frac{365-2}{365} = \frac{363}{365}$.

The probability of all 35 members of the group *not* having the same birth date is the product of these probabilities: $p = \frac{365}{365} \cdot \frac{365-1}{365} \cdot \frac{365-2}{365} \cdot \ldots \cdot \frac{365-34}{365}$.

Since the probability (q) that at least two members of the group *have* the same birth date, and the probability (p) that two members of the group do *not* have the same birth date is a certainty, the sum of those probabilities must, therefore, be 1. Thus, $p + q = 1$.

In this case, $q = 1 - \frac{365}{365} \cdot \frac{365-1}{365} \cdot \frac{365-2}{365} \cdot \ldots \cdot \frac{365-33}{365} \cdot \frac{365-34}{365} \approx .8143832388747152$. In other words, the probability that there will be a birth-date match in a randomly selected group of 35 people is somewhat greater than $\frac{8}{10}$. This is quite unexpected when one considers that there were 365 dates from which to choose. To further surprise you—or perhaps to shock you—we offer a list of probabilities that are correct and yet are hard to accept as such. (See figure 3.66.)

Number of People in Group	Probability of a Birth-Date Match	**Probability (in Percent) of a Birth-Date Match**
10	.116948	11.69%
15	.252901	25.29%
20	.411438	41.14%
25	.568700	56.87%
30	.706316	70.63%
35	.814383	81.44%
40	.891232	89.12%
45	.940976	94.10%
50	.970374	97.04%
55	.986262	98.63%
60	.994123	99.41%
65	.997683	99.77%
70	.999160	99.92%

Figure 3.66

Notice how quickly almost-certainty is reached. Were one to do this with the death dates of the first 35 presidents, one would notice that two died on March 8 (Millard Fillmore and William H. Taft) and three presidents died on the Fourth of July (John Adams, Jefferson, and Monroe). Above all, this demonstration should serve as an eye-opener about relying on intuition too much, and, at the same time, trusting probability even if it sometimes seems counterintuitive.

SOLUTION FOR PROBLEM 71

For those who insisted on using a calculator and avoided seeing the beauty of mathematics that leads to an elegant solution, we offer this result:

$$\sqrt[9]{9!} = \sqrt[9]{362,880} \approx 4.147166274,$$

$$\sqrt[10]{10!} = \sqrt[10]{3,628,800} \approx 4.528728688,$$

Therefore, $\sqrt[9]{9!} < \sqrt[10]{10!}$.

Using an algebraic method to establish the general case for this situation is to prove that $\sqrt[n+1]{(n+1)!} > \sqrt[n]{n!}$.

Although it is a bit tedious, we will provide this proof here. (Also, you can then compare it to the problem-solving strategy we shall offer as an alternative.)

Since $\sqrt[n+1]{n+1} > \sqrt[n+1]{n}$, $\sqrt[n+1]{n!(n+1)} > \sqrt[n+1]{n!n}$ (multiplying both sides of the inequality by $\sqrt[n+1]{n!}$); that is, $\sqrt[n+1]{(n+1)!} > \sqrt[n+1]{n!n}$.

Since $n^n > n!$ for $n > 1$, we have $n > \sqrt[n]{n!}$.

Therefore, $n!n > n!(n!)^{\frac{1}{n}} = (n!)^{\frac{n+1}{n}}$, and $\sqrt[n+1]{n!n} > \sqrt[n]{n!}$.

Thus, $\sqrt[n+1]{(n+1)!} > \sqrt[n+1]{n!n} > \sqrt[n]{n!}$.

As an alternative to this rather-difficult proof, we can work backward to solve this problem more elegantly! That is, we begin with what we want to establish. Let us start by taking both terms to a common power, namely, the ninetieth power:

$(\sqrt[9]{9!})^{90}$ <? $(\sqrt[10]{10!})^{90}$

$(9!)^{10}$ <? $(10!)^9$

$(9!)^9 \cdot 9!$ <? $(9!)^9 \cdot 10^9$

$9! < 10^9$

Which is actually $362,880 < 1,000,000,000$
Therefore, $\sqrt[9]{9!} < \sqrt[10]{10!}$.

SOLUTION FOR PROBLEM 72

Since each arc contains 30°, the points occur in diametrically opposite pairs. Now let's consider two of the distances called for in the problem, namely *PA* and *PG*. (See figure 3.67.)

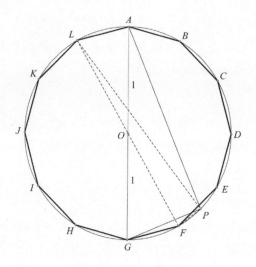

Figure 3.67

Since *AOG* is a diameter (of length 2), triangle *APG* is a right triangle. Applying the Pythagorean theorem:

$$PA^2 + PG^2 = AG^2 = 2^2 = 4.$$

Similarly, we conclude that since triangle *LPF* is a right triangle $PL^2 + PF^2 = LF^2 = 2^2 = 4$.

Organizing the data in this manner reveals six pairs of diametrically opposite points, and the sum of the squares of the distances from *P* is, therefore, $6 \cdot 4 = 24$.

SOLUTION FOR PROBLEM 73

There are many ways to approach the solution to this problem. Perhaps the most elegant way is to consider one of the six equilateral triangles as shown in figure 3.68. When the altitudes are drawn for each of these equilateral triangles, each equilateral triangle is divided into six equal-area right triangles. When we look at the area of the inner hexagon, whose area we seek to find, we notice that two of these six congruent right triangles are part of the shaded region, as shown in figure 3.69. In other words, one third of each of the six equilateral triangles comprises the area the shaded hexagon. Therefore, the inner hexagon is one third the area of the larger hexagon.

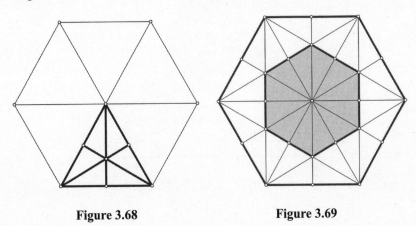

Figure 3.68   Figure 3.69

SOLUTION FOR PROBLEM 74

The nature of the problem would indicate that we should be using trigonometry to establish this triangle's side lengths. However, we can place this isosceles triangle within a square, since the altitude and the base have the same length. Then we draw the lines shown in figure 3.70, each of which is parallel to *EB* or *AC* and contains a midpoint or quarter point along the side of the square, thus making for the grid shown in the diagram.

With sides CE and BE having lengths 3 and 4, respectively, we can determine that the hypotenuse of the right triangle BCE is of length 5, and we have therefore shown the nature of this right triangle using a cleverly constructed diagram, and we have avoided lots of tedious computation. This is an example where we can see how an elegant solution underscores the beauty of mathematics.

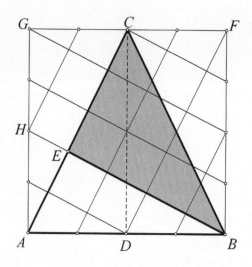

Figure 3.70

SOLUTION FOR PROBLEM 75

The traditional method for solving this problem is to evaluate each square, and then to add or subtract the appropriate terms (using a calculator, of course!). Some readers might elect to separate the series into two separate series to find the two separate sums, and then to combine them as shown below.

$$20^2 + 18^2 + 16^2 + \ldots + 4^2 + 2^2 \text{ and } -19^2 - 17^2 - 15^2 - \ldots - 3^2 - 1^2.$$

However, rather than to separate the given series into two different series as shown above, we can organize the data in the following way:
$$(20^2 - 19^2) + (18^2 - 17^2) + (16^2 - 15^2) + \ldots + (4^2 - 3^2) + (2^2 - 1^2).$$

Factoring each parenthetical expression, which shows the difference of two squares, we obtain:

$$(20 - 19)(20 + 19) + (18 - 17)(18 + 17) + (16 - 15)(16 + 15) + \ldots$$
$$+ (4 - 3)(4 + 3) + (2 - 1)(2 + 1).$$

For each of these terms, we observed that one of them will always have a value of 1. Thus giving us the following:

$$1 \cdot (20 + 19) + 1 \cdot (18 + 17) + 1 \cdot (16 + 15) + \ldots + 1 \cdot (4 + 3) + 1 \cdot (2 + 1).$$

What may have been a complicated-looking series at the start has now been reduced to merely finding the sum of the natural numbers 1 to 20, that is, $20 + 19 + 18 + 17 + \ldots + 4 + 3 + 2 + 1$.

Of course, we can just add these numbers to get our answer of 210, but it would be nice to add some more sophistication to the situation. We can use a method of adding consecutive numbers, which many mathematics teachers attribute to the famous German mathematician Carl Friedrich Gauss (1777–1855), who in his childhood is purported to have added the numbers 1 to 100 more quickly than anyone else in this class by simply adding the first and the last number, then adding the second and the next to last number, and so on, seeing each time that each pairing gave him the sum of 101. So all he had to do was multiply $50 \cdot 101$ to get his answer of 5,050.

We can employ this procedure here by pairing the numbers as shown below to get our required sum.

20 + 1 = 21

19 + 2 = 21

18 + 3 = 21

17 + 4 = 21

$$\vdots$$

11 + 10 = 21

We now have ten pairs whose sum is 21, giving us our required sum as: $10 \cdot 21 = 210$, also

$$20^2 - 19^2 + 18^2 - 17^2 + 16^2 - 15^2 + \ldots + 4^2 - 3^2 + 2^2 - 1^2 = 210.$$

SOLUTION FOR PROBLEM 76

This problem can benefit nicely from a strategy of considering the extreme case, which can also be considered as setting up the "worst-case scenario." This would have us be most "unlucky" and would have us pick the eight blue socks and the six green socks before a single black sock is selected. The next two picks would have to be black socks. In this situation, it took sixteen picks before we were *certain* of getting two black socks. Naturally, we might have achieved our goal of getting two black socks on our first two tries, but it was not guaranteed, and it would have been highly unusual. Even if we picked ten socks at random, we could not be certain that we had two black ones among them.

A simple extension of this problem can be illuminating. Now that we know it would take sixteen picks to be certain of getting at least two black socks, how many picks would be necessary to be certain of getting four black socks? If you know the answer immediately, you are a logical-reasoning whiz! Yes, you only need two more picks, for a total of eighteen. You've already accounted for the worst case, of picking all the socks of the other colors plus two black socks with your first sixteen picks, so the next two picks must provide you with only two black socks.

While still on the topic of making selections of socks, consider another variation of this issue. In a drawer, there are eight blue socks, six green socks, and twelve black socks. What is the smallest number of socks that must be taken from the drawer without looking at the socks to be certain of having two socks of the same color?

At first glance this problem appears to be similar to the previous problem. However, there is a slight difference. In this case, we are looking for a matching pair of socks of *any* color. We once again use the problem-solving strategy of considering the extreme case. The worst-case scenario has us picking one blue sock, one green sock, and one black sock in our first three picks. Thus, the fourth sock must provide us with a matching pair, regardless of what color it is. The smallest number of picks to guarantee a matching pair of socks is four.

SOLUTION FOR PROBLEM 77

Since we can feel the shoes and thereby determine right from left, we could take four right shoes from the closet. In that case, at least one of the shoes taken must be of a different color from the rest. Then, if we take a left-foot shoe from the closet, it must match at least one of the other right-foot shoes taken previously. Therefore, one must take five shoes from the closet to be certain of getting a right-foot shoe and left-foot shoe of the same color.

Using the same principle as we just did above, to solve the hair problem, we will label each person by the number of hairs on his or her head. By most estimates, the number of strands of hair on the average person's head is between 100,000 and 150,000. However, the most that anyone can have according to many sources is 200,000. So we will number up to as many as 200,000 persons. So, in a worst-case scenario, the first 200,000 people will all have a different number of hair strands on their head. In New York City, with a population of over eight million people, the 200,001st person will have to match one of the previous 200,000 already accounted for. Therefore, there will be at least two people in New York with the same number of hair strands on their head. (Obviously, with such a large population, there will be many other duplications.)

SOLUTION FOR PROBLEM 78

Now having eliminated the option of using a calculator, we seek a pattern among these fractions. One such pattern is as follows:

$$\frac{1}{1\cdot 2}+\frac{1}{2\cdot 3}+\frac{1}{3\cdot 4}+ \ldots +\frac{1}{49\cdot 50}.$$

We now inspect the partial sums to see if there is a pattern that may emerge.

$$\frac{1}{1\cdot 2}=\frac{1}{2}$$

$$\frac{1}{1\cdot 2}+\frac{1}{2\cdot 3}=\frac{2}{3}$$

$$\frac{1}{1\cdot 2}+\frac{1}{2\cdot 3}+\frac{1}{3\cdot 4}=\frac{3}{4}$$

$$\frac{1}{1\cdot 2}+\frac{1}{2\cdot 3}+\frac{1}{3\cdot 4}+\frac{1}{4\cdot 5}=\frac{4}{5}$$

By now you ought to notice a pattern, where each sum is equal to a fraction related in appearance to the last fraction being added. The two factors of the denominator of the last fraction in the sum seem to be used to determine the numerator and denominator of the sum. Therefore, the last fraction in our given series to be summed has factors of 49 and 50, which will determine the fraction representing the sum of the series.

$$\frac{1}{1\cdot 2}+\frac{1}{2\cdot 3}+\frac{1}{3\cdot 4}+ \ldots +\frac{1}{49\cdot 50} = \frac{49}{50}.$$

SOLUTION FOR PROBLEM 79

We begin by realizing the following:

$$\frac{1}{2} < \frac{2}{3},$$

$$\frac{3}{4} < \frac{4}{5},$$

$$\frac{5}{6} < \frac{6}{7},$$

$$\frac{7}{8} < \frac{8}{9}, \ldots,$$

$$\frac{99}{100} < \frac{100}{101}.$$

By multiplying these inequalities, we get the following:

$$\frac{1}{2} \cdot \frac{3}{4} \cdot \frac{5}{6} \cdot \frac{7}{8} \cdot \ldots \cdot \frac{99}{100} < \frac{2}{3} \cdot \frac{4}{5} \cdot \frac{6}{7} \cdot \frac{8}{9} \cdot \ldots \cdot \frac{100}{101}.$$

Then multiply both sides by $x = \frac{1}{2} \cdot \frac{3}{4} \cdot \frac{5}{6} \cdot \frac{7}{8} \cdot \ldots \cdot \frac{99}{100}$ to get

$$x^2 = \left(\frac{1}{2} \cdot \frac{3}{4} \cdot \frac{5}{6} \cdot \frac{7}{8} \cdot \ldots \cdot \frac{99}{100}\right)^2 < \left(\frac{1}{2} \cdot \frac{3}{4} \cdot \frac{5}{6} \cdot \frac{7}{8} \cdot \ldots \cdot \frac{99}{100}\right) \cdot \left(\frac{2}{3} \cdot \frac{4}{5} \cdot \frac{6}{7} \cdot \frac{8}{9} \cdot \ldots \cdot \frac{100}{101}\right)$$

$$= \frac{1}{\not{2}} \cdot \frac{\not{2}}{\not{3}} \cdot \frac{\not{3}}{\not{4}} \cdot \frac{\not{4}}{\not{5}} \cdot \frac{\not{5}}{\not{6}} \cdot \frac{\not{6}}{\not{7}} \cdot \frac{\not{7}}{\not{8}} \cdot \frac{\not{8}}{\not{9}} \cdot \ldots \cdot \frac{\not{99}}{\not{100}} \cdot \frac{\not{100}}{101} = \frac{1}{101},$$

With $x^2 < \frac{1}{101} < \frac{1}{100}$, we get $x < \frac{1}{10}$.

SOLUTION FOR PROBLEM 80

Many would begin by assuming that the hands would overlap at about 4:20, but after just a bit of thought, one should come to the realization that by 4:20, the hour hand will have moved away from the 4, so the overlap would occur sometime after 4:20. As a matter of fact, every twelve minutes, the hour hand moves one minute marker along the face of the clock. Therefore, between 4:12 and 4:24, the hour hand will be in the interval between 4:21 and 4:22. This problem can be dealt with in the same way that the typical algebra textbook treats uniform-motion problems, typically seen as one car overtaking another car. However, the distance here is measured in terms of minute markers and not in miles. The hour hand moves five minute markers per hour, while the minute hand moves sixty minute markers per hour. We need to find a number of minute markers that the minute hand must travel to overtake the hour hand. Let's call this required distance d. Therefore, the time required for the minute hand to meet the hour hand is its distance divided by its speed, or $\frac{d}{60}$. The distance required for the hour hand during that same time is $\frac{d-20}{5}$. Since the times are equal, we get the equation $\frac{d}{60} = \frac{d-20}{5}$. Then $d = \frac{12}{11} \cdot 20 = 21\frac{9}{11}$. Therefore, the hands will overlap at exactly $4:21\frac{9}{11}$.

Perhaps a simpler way of doing this is to use this fraction $\frac{12}{11}$ as a sort of "magic multiplier." That is, if we wish to find a desired position, say, after seven o'clock, one needs only to hold the hour hand stationary and allow the minute hand to travel to the desired position, which in this case is thirty-five minute markers. Then merely multiply this number by $\frac{12}{11}$ to get the exact overlap time, which in this case would be $7:38\frac{2}{11}$.

Using this scheme with the $\frac{12}{11}$, we find that the overlaps occur at $1:05\frac{5}{11}$, at $2:10\frac{10}{11}$, at $3:16\frac{4}{11}$, and so on. You should notice a pattern by now. To justify the $\frac{12}{11}$, think of the hands of the clock at noon. During the next twelve hours, the hour hand makes one revolution while the minute hand makes twelve revolutions. The minute hand coincides with the hour hand eleven times—including midnight but not noon, because we begin just after the noon moment. With the hands of the clock rotating at a uniform rate, the hands overlap every $\frac{12}{11}$ of an hour, or every $65\frac{5}{11}$ minutes.

You can create your own problems of this sort, such as finding when the hands of a clock after five o'clock will be perpendicular. All you need to do is to move the minute hand to the desired position, without the hour hand moving, and then use the correction factor of $\frac{12}{11}$. Multiplying the number of minutes that the minute hand moved by $\frac{12}{11}$, we get the exact time for the right angle—in this case $10 \cdot \frac{12}{11} = 10\frac{10}{11}$, which gives us the time for the right angle after five o'clock as $5:10\frac{10}{11}$. Applications of these clock problems should provide some fine entertainment.

SOLUTION FOR PROBLEM 81

During the twelve-hour period, the minute hand makes twelve complete revolutions and overlaps the hour hand 11 times. The times of overlap of these two hands are separated by $\frac{1}{11}$ of the circumference of the clock. Using the same reasoning, we find that the minute hand and the second hand overlap at fifty-nine locations on the circumference of the clock. Since the numbers eleven and fifty-nine are prime numbers, they have no common divisor. Therefore, the three hands of the clock will never overlap except at the beginning and end—twelve o'clock.

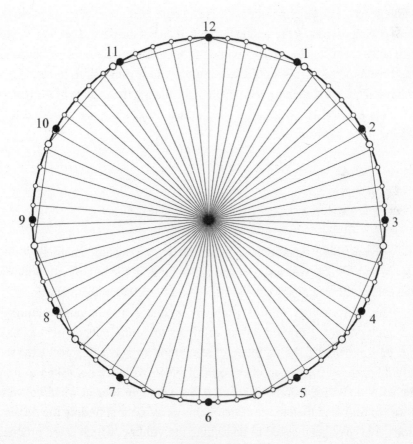

Figure 3.71

The regular eleven-sided polygon (with the vertices indicated by large white circles in figure 3.71) shows the overlapping points of the hour and minute hand. The small white dots are the vertices of a regular fifty-nine-sided polygon. There you can see that there are no overlaps of all three hands.

SOLUTION FOR PROBLEM 82

We have been taught that in order to find the volume of a box, we need to know the dimensions of the box—that is, we need to know the lengths of the width, length, and height of the box. This would lead us to set up a system of three equations, each representing the area of one of the three faces. We would then solve them simultaneously to find the three dimensions. Having obtained the three dimensions, we would then multiply these dimensions to obtain the volume. This is a straightforward method for finding the volume of the box.

This method could be a bit time-consuming, since we have the following three equations:

$$w \cdot l = 165$$
$$w \cdot h = 176$$
$$l \cdot h = 540$$

Solving these equations simultaneously, we would obtain $w = \frac{22}{3}$, $l = \frac{45}{2}$, and $h = 24$. We can calculate the volume of the box by multiplying these three dimensions: $w \cdot l \cdot h = \frac{22}{3} \cdot \frac{45}{2} \cdot 24 = 3{,}960$ cubic inches.

The curious part of this problem is that our previous training—although correct —often does not allow us to "think outside of the box" (no pun intended!). As an alternative method to solve this problem, we might approach it from another point of view. We were not asked to find the individual dimensions, w, l, and h. So we might look at what we were asked to find, and then to see if that helps us avoid first finding the dimensions. We have been asked to determine the volume, which is the product of the three dimensions.

$V = w \cdot l \cdot h$

$V^2 = (w \cdot l \cdot h) \cdot (w \cdot l \cdot h)$

$V^2 = (w \cdot l) \cdot (w \cdot h) \cdot (l \cdot h)$

We now have a formula that asks us to provide the very information we were originally given, so the problem becomes rather simple.

$V^2 = 165 \cdot 176 \cdot 540 = (3 \cdot 5 \cdot 11) \cdot (2^4 \cdot 11) \cdot (2^2 \cdot 3^3 \cdot 2^4 \cdot 5) = 2^6 \cdot 3^4 \cdot 5^2 \cdot 11^2$

$V = \sqrt{2^6 \cdot 3^4 \cdot 5^2 \cdot 11^2} = 2^3 \cdot 3^2 \cdot 5 \cdot 11 = 9 \cdot 8 \cdot 5 \cdot 11 = 99 \cdot 8 \cdot 5 = 792 \cdot 5$

$= 792 \cdot \dfrac{10}{2} = 7{,}920 \div 2 = 3{,}960$ cubic inches.

SOLUTION FOR PROBLEM 83

To solve this problem, we need only to continuously multiply by four as follows:

Multiplying by four $\frac{1}{4}\left\{\frac{1}{4}\left[\frac{1}{4}\left(\frac{1}{4}x-\frac{1}{4}\right)-\frac{1}{4}\right]-\frac{1}{4}\right\}-\frac{1}{4}=0$

gives us $1\left\{\frac{1}{4}\left[\frac{1}{4}\left(\frac{1}{4}x-\frac{1}{4}\right)-\frac{1}{4}\right]-\frac{1}{4}\right\}-1=0$.

Then, continuously multiplying by four, we get:

$1\left[\frac{1}{4}\left(\frac{1}{4}x-\frac{1}{4}\right)-\frac{1}{4}\right]-1-4=0$

$1\left(\frac{1}{4}x-\frac{1}{4}\right)-1-4-16=0$

$x-1-4-16-64=0$

$x-85=0$

Therefore, $x = 85$.

This looked overwhelming at the start, but systematically approaching it made the problem simple!

SOLUTION FOR PROBLEM 84

Follow along, as the steps to a conclusion speak for themselves, while we pair the terms and evaluate them as follows:

$\tan 15° \cdot \tan 75° = \tan 15° \cdot \tan (90° - 15°) = \tan 15° \cdot \cot 15° = \tan 15° \cdot \frac{1}{\tan 15°} = 1$, $\tan 30° \cdot \tan 60° = \tan 30° \cdot \tan (90° - 30°) = \tan 30° \cdot \cot 30° = \tan 30° \cdot \frac{1}{\tan 30°} = 1$, $\tan 45° = 1$.

Therefore, $\tan 15° \cdot \tan 30° \cdot \tan 45° \cdot \tan 60° \cdot \tan 75°$

$= (\tan 15° \cdot \tan 75°) \cdot (\tan 30° \cdot \tan 60°) \cdot \tan 45° = 1 \cdot 1 \cdot 1 = 1$.

Problem solved—by thoughtful pairing.

SOLUTION TO PROBLEM 85

Most people when faced with this question respond with the answer $\frac{1}{8}$, since there are eight sides to the octagon. This is a wrong answer!

The correct answer is $\frac{1}{4}$.

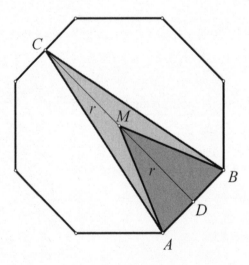

Figure 3.72

Rather than to draw all sorts of auxiliary lines, this answer can be elegantly obtained by considering the triangle AMB (see figure 3.72), which is clearly $\frac{1}{8}$ of the area of the regular octagon. Where the altitude of triangle *AMB* has length *r*, we can easily see that the altitude of triangle *ACB* has length 2*r*, and since both triangles have the same base, *AB*, we can conclude that the area of triangle *ACB* is twice that of triangle *AMB*, or $\frac{1}{4}$ of the area of the octagon.

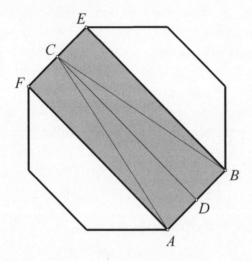

Figure 3.73

The rectangle formed as shown in figure 3.73 has double the area of triangle *ABC*, since it has the same base *AB* and the same altitude *CD*. Therefore, the rectangle has half the area of the octagon.

SOLUTION TO PROBLEM 86

Through some inspection you may notice that there are some familiar patterns within this sequence, namely the powers of 2 and the powers of 3, such as:

- 1, 2, 4, 8, 16, 32, 64, . . .
- 1, 3, 9, 27, 81, . . .

Yet there are other numbers included in this sequence that are not included in these lists of powers, such as 6, 12, 18, and others. With a bit of thought, we will notice that they are all products of powers of 2 and powers of 3 of the form: $2^m \cdot 3^n$, where $m, n > 0$. By the way, such numbers have been tagged with the name "3-smooth numbers" as they are numbers where none of their prime factors is greater than 3.

An integer is a *k-smooth number* if it has no prime factors greater than k.

For those who would like to play a trick on their friends, you might ask for the missing terms in the following sequence: 4, 14, 34, 42, 59, 125, ____, 168, 175, 181, ___, 200, 207. Folks struggle over this, not realizing that this is merely a trick question that a seasoned New Yorker should be able to recognize. These are the subway stops along the A-train, which runs along the west side of Manhattan Island. We mention this because of the famous jazz song "Take the A-Train," which was written by Billy Strayhorn and was the signature tune of the Duke Ellington orchestra.

SOLUTION TO PROBLEM 87

When this question is posed, a natural reaction is to try some pattern placements of the dominoes. Before long, frustration begins to set in since this approach cannot be successful.

A careful reading of the question reveals that it does not say to *do* this tile covering; it asks if it can be done. Yet, because of the way we have been trained, the question is often misread and interpreted as "do it."

A bit of clever insight helps. Ask yourself the question: "When a domino tile is placed on the chessboard, what sort of squares are covered?" A black square and a white square must be covered by each domino placed on the chessboard. Are there an equal number of black and white squares on the truncated chessboard? No! There are two fewer black squares than there are white squares. Therefore, it is impossible to cover the truncated chessboard with the thirty-one domino tiles, since there must be an equal number of black and white squares.

SOLUTION FOR PROBLEM 88

To consider all possible combinations for exiting the cave under the given conditions can be very time-consuming. Therefore, it is advisable to cleverly organize the data to develop a proper strategy.

This problem is only solvable in the twelve hours allotted if persons A and B walk together, since this covers the greatest time period. This minimal amount of time will only be reached if the two slowest persons walk together. (See figure 3.74.)

Cave	Travelers	At the Exit	Time
A, B, C, D	C, D	C, D	2
A, B, D	D	C, D	1
A, B, D	A, B	A, B, C	5
C, D	C	A, B, C	2
C, D	C, D	C, D	2
			12

Figure 3.74

If we replace the people with their travel time, we can have the following overview (see figure 3.75).

Cave	Travelers	At the Exit	Time
1, 2, 4, 5	–	–	–
4, 5	1, 2 →	–	2
4, 5	–	1, 2	–
4, 5	1 ←	2	1
1, 4, 5	–	2	–
1	4, 5 →	2	5
1	–	2, 4, 5	–
1	2 ←	4, 5	2
1, 2	–	4, 5	–
–	1, 2 →	4, 5	2
–	–	1, 2, 4, 5	–
			12

Figure 3.75

SOLUTION FOR PROBLEM 89

The peasant's way to solve this problem is to list all the possible numbers that can be formed with the digits of the number 975. They are: 579, 597, 759, 795, 957, and 975. Then all one needs to do is to add the numbers. Their sum is 4,662.

The poet's curious way to approach this problem is to note that each of the digits must appear twice in each of the place-value positions. Therefore, the sum of the digits in the sum must be 21 $(9 + 7 + 5)$ in each place. We can, therefore, obtain the required sum by getting:

$$579 + 597 + 759 + 795 + 957 + 975$$
$$= (5 \cdot 100 + 7 \cdot 10 + 9) + (5 \cdot 100 + 9 \cdot 10 + 7) + (7 \cdot 100 + 5 \cdot 10 + 9)$$
$$+ (7 \cdot 100 + 9 \cdot 10 + 5) + (9 \cdot 100 + 5 \cdot 10 + 7) + (9 \cdot 100 + 7 \cdot 10 + 5)$$
$$= (5 \cdot 100 + 5 \cdot 100 + 7 \cdot 100 + 7 \cdot 100 + 9 \cdot 100 + 9 \cdot 100)$$
$$+ (7 \cdot 10 + 9 \cdot 10 + 5 \cdot 10 + 9 \cdot 10 + 5 \cdot 10 + 7 \cdot 10)$$
$$+ (9 + 7 + 5 + 9 + 5 + 7 + 5)$$
$$= 2 \cdot (5 + 7 + 9) \cdot 100 + 2 \cdot (5 + 7 + 9) \cdot 10 + 2 \cdot (5 + 7 + 9)$$
$$= 42 \cdot 100 + 42 \cdot 10 + 42 = 4,662.$$

SOLUTION FOR PROBLEM 90

We provided this last problem to show that there is another dimension to logical thinking than arithmetic or algebraic keys to be searched for. The algorithm here simply assigns to each digit a number based on the number of closed circuits that the shape of the digit demonstrates. You will notice, for example, that the digit 8 exhibits two close loops, hence, we assign the number 2 to it. Silly as this may seem, this sort of algorithm has been used in some cryptography. Figure 3.76 shows the numbers assigned to each of the digits.

The numerals	0	1	2	3	4	5	6	7	8	9
Number of closed circuits	1	0	0	0	0	0	1	0	2	1

Figure 3.76

Although some may argue that the digit 4 demonstrates a closed loop, we excluded it because the closure is comprised of linear components and not circular ones.

Chapter 4

MEAN CURIOSITIES

Comparing Measures of Central Tendency —from a Geometric Point of View

In mathematics and in statistics we frequently use measures of central tendency—that is, we use various means such as the *arithmetic mean* (in common terms: the average), the *geometric mean*, and the *harmonic mean*. Our knowledge of them has been traced back to ancient times. As a matter of fact, the historian Iamblichos of Chalkis (ca. 250–330 CE) reported that Pythagoras (ca. 580/560–ca. 496/480 BCE), after a visit to Mesopotamia, brought back to his followers a knowledge of these three measures of central tendency. This may be one reason why today they are often referred to as *Pythagorean means*. We tend to use these means in statistical analyses, but there are some rather-enlightening views when we inspect them and compare them geometrically.

Let's begin by introducing these three means for two values a and b as follows[1]:

- The *arithmetic mean* is $AM(a, b) = a \Ⓐ\ b = \dfrac{a+b}{2}$,

- The *geometric mean* is $GM(a, b) = a \Ⓖ\ b = \sqrt{a \cdot b}$, and

- The *harmonic mean* is $HM(a, b) = a \Ⓗ\ b = \dfrac{2}{\dfrac{1}{a}+\dfrac{1}{b}} = \dfrac{2ab}{a+b}$.

For convenience throughout this discussion we will use these representations as we discuss them in greater detail:

THE ARITHMETIC MEAN

Before comparing the relative magnitude of these measures of central tendency, or means, we ought to see what they actually represent. The arithmetic mean is simply the commonly used "average" of the data being considered—that is, the sum of the data divided by the number of data items included in the sum. In a simple example, if we want to find the average—or arithmetic mean—between the two values of 30 and 60, we take their sum, 90, and divide it by 2 to get 45.

We can also see the arithmetic mean as taking us to the middle of an arithmetic sequence—that is, a sequence with a common difference between terms—such as 2, 4, 6, 8, 10. To get the arithmetic mean, we divide the sum by the number of numbers being averaged. Here we have: $\frac{2+4+6+8+10}{5} = \frac{30}{5} = 6$, which, as we expected, happens to be the middle number in the sequence of an odd number of values.

THE HARMONIC MEAN

If we take an arithmetic sequence such as $1, 2, 3, 4, 5$, and take the reciprocals, we have a harmonic sequence: $1, \frac{1}{2}, \frac{1}{3}, \frac{1}{4}, \frac{1}{5}$. We can tie the harmonic mean to the arithmetic mean by simply indicating that the harmonic mean is the reciprocal of the arithmetic mean of the reciprocals of the numbers. We would do this step by step, as follows:

To get the harmonic mean $HM(1, 2, 3, 4, 5)$ of a given sequence $1, 2, 3, 4, 5$, we first find the arithmetic mean $AM\left(1, \frac{1}{2}, \frac{1}{3}, \frac{1}{4}, \frac{1}{5}\right)$ of the reciprocals of the sequence, that is,

$$AM\left(1, \frac{1}{2}, \frac{1}{3}, \frac{1}{4}, \frac{1}{5}\right) = \frac{1 + \frac{1}{2} + \frac{1}{3} + \frac{1}{4} + \frac{1}{5}}{5} = \frac{\frac{60+30+20+15+12}{60}}{5} = \frac{\frac{137}{60}}{5} = \frac{137}{300} (\approx 0.457).$$

We then take the reciprocal of this value to get the harmonic mean,

$$HM(1, 2, 3, 4, 5) = \frac{300}{137} (\approx 2.19).$$

Another way of looking at the same procedure is the following:

$$HM(1, 2, 3, 4, 5) = \frac{1}{AM\left(1, \frac{1}{2}, \frac{1}{3}, \frac{1}{4}, \frac{1}{5}\right)} = \frac{1}{\frac{1 + \frac{1}{2} + \frac{1}{3} + \frac{1}{4} + \frac{1}{5}}{5}} = \frac{5}{1 + \frac{1}{2} + \frac{1}{3} + \frac{1}{4} + \frac{1}{5}} = \frac{300}{137} (\approx 2.19).$$

Where the harmonic mean is particularly useful is to find the average of rates over a common base. For example, consider the question of finding the average speed of a round-trip journey. Suppose you were going at a rate of 30 mph and returning over the same route (the base) at 60 mph. One might be tempted to simply find the arithmetic mean, $\frac{30 + 60}{2} = 45$. This would be incorrect. Is it fair to give equal value to the 30 mph trip as to the 60 mph trip when the former took twice as long as the latter? Here we would invoke the harmonic mean, which would require us to get the reciprocal of the arithmetic mean of the reciprocals of the two numbers. That is, for the harmonic mean of 30 and 60, we get $\frac{1}{\frac{\frac{1}{30} + \frac{1}{60}}{2}} = \frac{2}{\frac{1}{30} + \frac{1}{60}} = \frac{2}{\frac{3}{60}} = \frac{120}{3} = 40$.

This could, of course, be more simply done by using the formula $\frac{1}{\frac{\frac{1}{a} + \frac{1}{b}}{2}} = \frac{2ab}{a + b}$.

THE GEOMETRIC MEAN

The geometric mean gets its name from a simple interpretation in geometry. A rather-common application of the geometric mean is obtained when we consider the altitude to the hypotenuse of a right triangle. In figure 4.1, *CD* is the altitude to the hypotenuse of right triangle *ABC*.

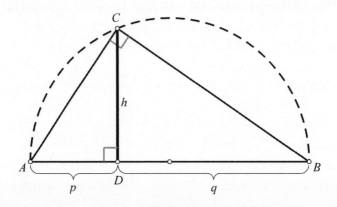

Figure 4.1

From the triangle similarity, $\Delta ADC \sim \Delta CDB$, we get $\frac{AD}{CD} = \frac{CD}{BD}$, or $\frac{p}{h} = \frac{h}{q}$. This then gives us $h = \sqrt{pq}$, which has h as the geometric mean between p and q.

The geometric mean is also the "middle" of a geometric sequence. Take, for example, the geometric sequence[2] 2, 4, 8, 16, 32. To get the geometric mean of these five numbers, we find the fifth root of their product: $\sqrt[5]{2 \cdot 4 \cdot 8 \cdot 16 \cdot 32} = \sqrt[5]{32{,}768} = 8$, which is the middle number. Again, the sequence would have an odd number of numbers in order to have a middle number.

Before we present some unusual geometric methods for comparing the magnitude of these three means, we shall show how these three means may be compared in size using simple algebra.

For the two non-negative numbers a and b,

$$(a-b)^2 \geq 0$$
$$a^2 - 2ab + b^2 \geq 0$$

Add $4ab$ to both sides:

$$a^2 + 2ab + b^2 \geq 4ab$$
$$(a+b)^2 \geq 4ab$$

Take the positive square root of both sides:

$$a+b \geq 2\sqrt{ab}$$

$$\text{or} \quad \frac{a+b}{2} \geq \sqrt{ab}$$

This implies that the *arithmetic mean* of the two numbers a and b is greater than or equal to the *geometric mean*. (Equality is true only if $a = b$.)

Beginning as we did above, and then continuing along as shown below, we get our next desired result: a comparison of the geometric mean and the harmonic mean.

For the two non-negative numbers a and b,

$$(a-b)^2 \geq 0$$
$$a^2 - 2ab + b^2 \geq 0$$

Add $4ab$ to both sides:
$$a^2 + 2ab + b^2 \geq 4ab$$
$$(a+b)^2 \geq 4ab$$

Multiply both sides by ab:

$$ab(a+b)^2 \geq 4a^2b^2$$

Divide both sides by $(a + b)^2$:
$$ab \geq \frac{4a^2b^2}{(a+b)^2}$$

Take the positive square root of both sides:

$$\sqrt{a \cdot b} \geq \frac{2ab}{a+b}$$

This tells us that the *geometric mean* of the two numbers a and b is greater than or equal to the *harmonic mean*. (Here, equality holds whenever one of these numbers is zero, or if $a = b$).

We can, therefore, conclude that

arithmetic mean $\geq$ *geometric mean* $\geq$ *harmonic mean.*

Comparing the Three Means Geometrically—
Using a Right-Angled Triangle

The comparison of the three means in terms of their relative size was known to the ancient Greeks, as we find in the writings of Pappus of Alexandria (ca. 250–350 CE). We will now embark on a geometric journey to consider various ways that the relative sizes of these means can be compared using simple geometric relationships.

In figure 4.2, we have a right triangle with an altitude drawn to the hypotenuse, where the hypotenuse is partitioned by the altitude into two segments of lengths a and b, and $a \leq b$. Here we show the line segments that can represent the three means, and where we can then "see" their relative sizes. That is, $a \circledH b \leq a \circledG b \leq a \circledA b$. We see this again in figure 4.3, where the right triangles once again came into play.

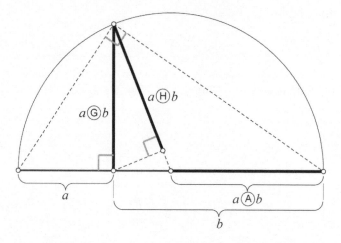

Figure 4.2

To justify our visual observation, we begin by considering figure 4.3, where we notice that CE is a leg of right triangle CED; therefore, $CE \leq CD$. Since the radius of the circumscribed circle of triangle ABC is longer than the altitude to the hypotenuse of the triangle ABC, we have that $CD \leq MB$. Combining these inequalities gives us that $CE \leq CD \leq MB$. Our task is to show that these three segments, whose relative lengths we have established, actually represent the three means of a and b as we earlier stated.

First, we know that AB is the diameter of the circle with center at M and radii $MA = MB = MC = \dfrac{a+b}{2}$, which is the arithmetic mean between a and b.

To find the geometric mean of a and b we begin with the altitude, CD, to the hypotenuse of right triangle ABC, which partitions the right triangle into two similar triangles $(\triangle ADC \sim \triangle CDB)$ and therefore, $\dfrac{a}{CD} = \dfrac{CD}{b}$, which leads to $CD^2 = a \cdot b$; thus, $CD = \sqrt{a \cdot b}$, which is the geometric mean of a and b.

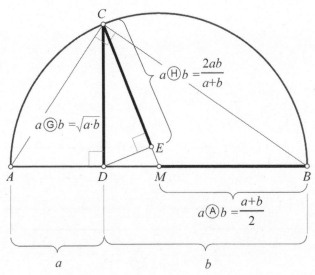

Figure 4.3

From the similar triangles $(\triangle CDM \sim \triangle ECD)$ located in right triangle CDM, we can get $CD^2 = MC \cdot CE$, which yields $CE = \dfrac{CD^2}{CM} = \dfrac{a \cdot b}{\frac{a+b}{2}} = \dfrac{2ab}{a+b}$, which is the harmonic mean. Having now justified how the line segments, which we size-ordered as $CE \le CD \le MB$, can represent the various means of a and b, we have therefore shown geometrically that $\dfrac{2ab}{a+b} \le \sqrt{a \cdot b} \le \dfrac{2ab}{a+b}$.

Some may notice that $a \cdot b = \dfrac{a+b}{2} \cdot \dfrac{2ab}{a+b}$, therefore, we can establish another relationship that ties the three means together:

$$GM(a, b)^2 = AM(a, b) \cdot HM(a, b),$$

or written another way: (a Ⓖ b)^2 = (a Ⓐ b) · (a Ⓗ b).

With some further simple algebraic manipulation, we can get the following relationship:

$$\frac{a}{\frac{a+b}{2}} = \frac{2a}{a+b} = \frac{2a \cdot b}{(a+b) \cdot b} = \frac{2ab}{(a+b)} \cdot \frac{1}{b} = \frac{2ab}{\frac{a+b}{b}}, \text{ or in short, } \frac{a}{\frac{a+b}{2}} = \frac{\frac{2ab}{a+b}}{b}, \text{ which}$$

then again shows us that $\dfrac{a}{AM(a,b)} = \dfrac{HM(a,b)}{b}$. We show this relationship geometrically in figure 4.4.

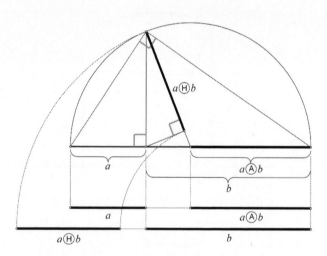

Figure 4.4

Clearly, these means can be applied to more than two values. For example, when we find these means for three numbers we have the following:

$$AM(a,\, b,\, c) = \frac{a+b+c}{3},$$

$$GM(a,\, b,\, c) = \sqrt[3]{a \cdot b \cdot c}, \text{ and}$$

$$HM(a,\, b,\, c) = \frac{3}{\frac{1}{a}+\frac{1}{b}+\frac{1}{c}} = \frac{3abc}{ab+ac+bc}.$$

The expected extrapolation would follow for finding the various means of any larger number of items.

Comparing the Three Means Geometrically—Using a Rectangle

Some geometric comparisons of the means exhibit curious ways in which we can use geometry to explain mathematical comparisons. For example, if we want more evidence that the geometric mean is always less than or equal to the arithmetic mean, we refer to figure 4.5, where we have a rectangle with side lengths a and b, and a square with sides of length $\frac{a+b}{2}$ (which is the arithmetic mean of a and b).

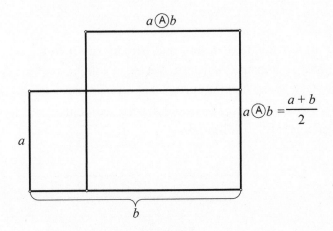

Figure 4.5

The area of the rectangle is $A_R = a \cdot b$, while the area of the square is as $A_s = \left(\frac{a+b}{2}\right)^2$. It has been demonstrated that when a square and a rectangle have the same perimeter—in this case, $2(a+b)$—the area of a square is always greater than the area of the rectangle. Therefore, $\left(\frac{a+b}{2}\right)^2 \geq a \cdot b$. Now, if we take the square root of each side of the equation, we get: $\sqrt{a \cdot b} \geq \frac{a+b}{2}$, which is what we wanted to show—namely, that the geometric mean is less than the arithmetic mean.

Having now compared the three most common means, we now introduce a fourth mean.

This new measure of central tendency, or mean, is called the _root-mean-square_,[3] which we can represent as $RMS(a, b) = a \circledR b = \sqrt{\frac{a^2+b^2}{2}}$. Statisticians have developed this other method of finding central tendency of numbers that permits negative numbers to be included. To ignore the negative aspects of these numbers, each of the numbers is squared and the square root of the arithmetic mean of these squares is taken. To get the root-mean-square of the sequence of numbers –10, –4, –3, 2, 5, 7, 9, we do the following:

$$\sqrt{\frac{(-10)^2+(-4)^2+(-3)^2+2^2+5^2+7^2+9^2}{7}} = \sqrt{\frac{284}{7}} \approx 6.37.$$

Let's inspect this new mean—that is, compare it to the other three means—from a geometric viewpoint. We begin with the root-mean-square of the two values, $RMS(a, b) = a \circledR b = \sqrt{\frac{a^2+b^2}{2}}$ and compare it to the previously defined means. We would expect and hope it will lead us to some surprising results. Follow along.

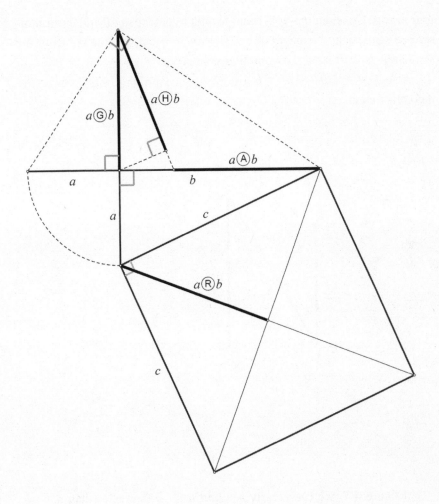

Figure 4.6

Referring to figure 4.6, we construct a square with the hypotenuse length c as a side length. Applying the Pythagorean theorem to this right triangle, we get $c = \sqrt{a^2+b^2}$. The diagonal of the square is $d = \sqrt{c^2+c^2}$ $= c\sqrt{2}$. Then one-half the length of the diagonal is $\frac{d}{2} = \sqrt{2} \cdot \sqrt{a^2+b^2} = \frac{\sqrt{2}}{\sqrt{2} \cdot \sqrt{2}} \cdot$ $\sqrt{a^2+b^2} = \sqrt{\frac{a^2+b^2}{2}}$, which is the root-mean-square of a and b.

Now that we have identified a line segment of the length of the root-mean-square, we will now begin to compare the relative magnitudes of the four means. Consider the right triangle with hypotenuse AB and its circum-scribed semicircle. We have $AB = AD + DB = a + b$, and the radius of the semicircle is $a ⓐ b$, as shown in figure 4.7.

Also $DM = DB - MB = b - \frac{a+b}{2} = \frac{b-a}{2}$. Applying the Pythagorean theorem we get

$$\sqrt{\left(\frac{a+b}{2}\right)^2 + \left(\frac{b-a}{2}\right)^2} = \sqrt{\frac{a^2 + 2ab + b^2 + b^2 - 2ab + a^2}{4}} = \sqrt{\frac{2a^2 + 2b^2}{4}} = \sqrt{\frac{a^2 + b^2}{2}} = RMS(a,\ b)$$
$$= a ⓡ b.$$

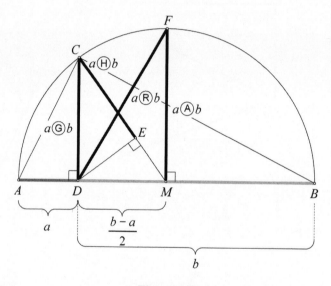

Figure 4.7

In figure 4.7, we can clearly see that $a ⓡ b$ is greater than $a ⓐ b$, which is greater than $a ⓖ b$, which in turn is greater than $a ⓗ b$. That is, $FD > FM > CD > CE$.

Or symbolically, we have: $a ⓡ b > a ⓐ b > a ⓖ b > a ⓗ b$. Only when $a = b$ are the means equal.

These geometric comparisons give visual meaning to what otherwise is more abstract and perhaps less convincing.

AN ALTERNATIVE VIEWPOINT

It is often said that a picture is worth a thousand words. In the next several figures, we can essentially repeat the previous means comparisons by just moving a rectangle and a square to various positions, showing in another fashion the previous comparisons. This was presented by Eli Maor[4] (1938–) in a very clever and clear fashion. We leave it without words, as the figures alone should tell the story.

We begin here with two quadrilaterals having the same perimeter $2a + 2b$. In order to get the geometric interpretation of the *arithmetic mean* from the rectangle with sides of length a and b, we use figure 4.8, which shows a square with sides of length $x = a \, Ⓐ \, b = \frac{a+b}{2}$, that is, the arithmetic mean.

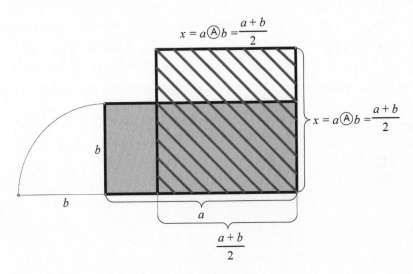

Figure 4.8

The *geometric mean* is obtained from the rectangle with sides lengths a and b and occurs as the side lengths $x = a \, Ⓖ \, b = \sqrt{a \cdot b}$ of a different square, whose area equals that of the rectangle shown in figure 4.9. From this diagram, we can clearly see that the arithmetic mean $\frac{a+b}{2}$ is greater than the geometric mean $\sqrt{a \cdot b}$, since the geometric mean is clearly shorter than the radius of the large semicircle.

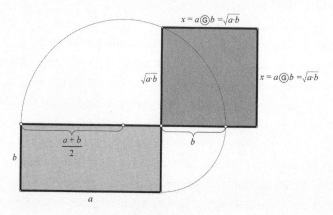

Figure 4.9

The *harmonic mean* in figure 4.10 is the side length of the square that results from the rectangle sides of lengths a and b. The square has side length $x = a \oplus b = \frac{2}{\frac{1}{a} + \frac{1}{b}} = \frac{2ab}{a+b}$. Once again, the diagram allows us to see very clearly that the harmonic mean, $\frac{2ab}{a+b}$, which is represented by the square's side length, is shorter than the radius of the large semicircle, which represents the arithmetic mean, $\frac{a+b}{2}$.

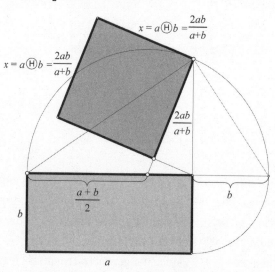

Figure 4.10

The *root-mean-square* is shown as the side length of the square in figure 4.11, again based on the dimensions of the rectangle with side lengths a and b. The sides of the square are of length $x = a \circledR b = \sqrt{\frac{a^2+b^2}{2}}$, which is the root-mean-square of a and b. We can see that the side of the square in figure 4.11 is also the hypotenuse of a right triangle where one leg is the radius of the large semicircle, which just happens to be the arithmetic mean of a and b. Therefore, figure 4.11 (with $d_S = d_R$), shows us that the root-mean-square is larger than the arithmetic mean.

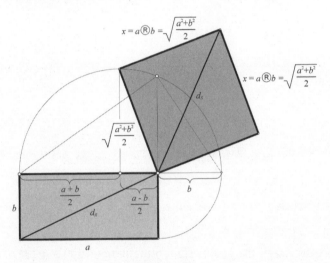

Figure 4.11

Comparing Means Using a Hyperbola

We began with the Cartesian plane with the help of the positive nape of a hyperbola[5] whose equation is $y = \frac{1}{x}$. (See figure 4.12.) We are particularly interested in the two points on the hyperbola: $A(a, \frac{1}{a})$ and $B(b, \frac{1}{b})$, where $a < b$.

With M as the midpoint of segment AB, we get $x_M = \frac{a+b}{2} = a \circledA b = AM(a, b)$, which is the arithmetic mean between a and b.[6] Similarly, along the y-axis we get $y_M = \frac{\frac{1}{a}+\frac{1}{b}}{2} = \frac{a+b}{2ab} = \frac{1}{HM(a,b)}$.

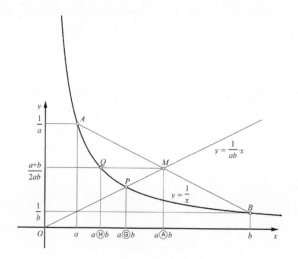

Figure 4.12

Consider the line OM, and $\dfrac{y_M}{x_M} = \dfrac{\frac{a+b}{2ab}}{\frac{a+b}{2}} = \dfrac{1}{ab}$, we get the equation for OM to be $y = \dfrac{1}{ab} \cdot x$. Notice that OM intersects the hyperbola $y = \dfrac{1}{x}$ at point P, so that because $\dfrac{1}{ab} \cdot x = \dfrac{1}{x}$, we get $x_P = \sqrt{ab}$, which is the geometric mean $GM(a, b) = a \circledcirc b$. (Note that $y_P = \dfrac{1}{\sqrt{ab}}$.)

As we now seek the geometric representation (figure 4.12) of the harmonic mean, we notice that the line through point M and parallel to the x-axis intersects the hyperbola $y = \dfrac{1}{x}$ at point Q. This determines $y_Q = y_M = \dfrac{a+b}{2ab}$, and thus, $y_Q = \dfrac{a+b}{2ab} = \dfrac{1}{x_Q}$, which then can be written as $x_Q = \dfrac{2ab}{a+b}$. This is the harmonic mean between a and b: $HM(a, b) = a \oplus b$.

To pursue the geometric interpretation of the root-mean-square, $RMS(a, b) = a \circledR b = \sqrt{\dfrac{a^2+b^2}{2}}$, we will refer to figure 4.13, where R is the point of intersection of the circle with center at O and radius length $a \circledA b = \dfrac{a+b}{2}$ and the line parallel to the y-axis and containing point P, which is the line $x = a \circledcirc b = \sqrt{ab}$. The line parallel to the x-axis through the point R intersects the parallel line to the y-axis, which contains the point M (that is, the line $x = a \circledA b = \dfrac{a+b}{2}$) at the point S.

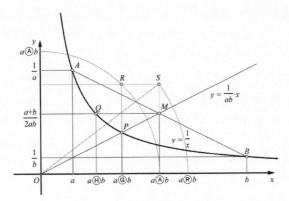

Figure 4.13

The circle with center O and radius $OS = a \circledA b = \dfrac{a+b}{2}$ intersects the x-axis to determine the root-mean-square $a \circledR b = \sqrt{\dfrac{a^2+b^2}{2}}$.[7]

Comparing Means Using a Trapezoid

Now that we have a visual feel for these means, we can search for other geometric evidence for our now-established comparison of the four means. Consider trapezoid $ABCD$[8] with bases a and b (where $a \geq b$) and median m_A joining side midpoints E and F, as shown in figure 4.14. This immediately gives us the length EF as the arithmetic mean $a \circledA b = \dfrac{a+b}{2}$, which we can justify in the following way.

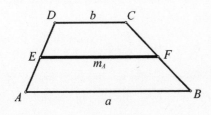

Figure 4.14

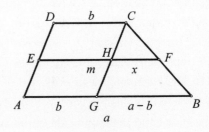

Figure 4.15

In figure 4.15, we have $CG \parallel AD$, $BF = CF$, $x = FH$ and $a - b = BG$. We then get $\frac{FH}{BG} = \frac{CF}{BC} = \frac{CF}{2 \cdot CF} = \frac{1}{2}$, therefore, $x = \frac{1}{2} \cdot (a - b)$, which yields $m = EF = EH + HF = b + x = b \cdot \frac{1}{2}(a - b) = \frac{a+b}{2} = a \text{ Ⓐ } b$.

Thus, we have shown how we got $m_A = EF$ as the arithmetic mean between the two parallel bases.

In order to determine the geometric mean $a \text{ Ⓖ } b = \sqrt{a \cdot b}$ on the trapezoid $ABCD$ with base lengths a and b, we draw line EF parallel to the bases in such a way that the trapezoids $ABEF$ and $FECD$ are similar so that $\frac{AB}{EF} = \frac{EF}{CD}$. (See figure 4.16.)

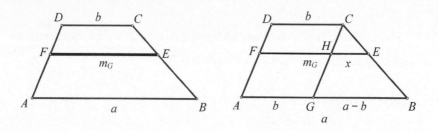

Figure 4.16   **Figure 4.17**

In figure 4.17, we add an auxiliary line $CG \parallel AD$, where $x = EH$, $x + b = EH + HF = EF$ and $a - b = BG$. From the similarity of the trapezoids $ABEF$ and $FECD$, we get $\frac{AB}{EF} = \frac{EF}{CD}$, or $\frac{a}{x+b} = \frac{x+b}{b}$. Then, by multiplying by $b(x + b)$, we find that $ab = (x + b)^2$, and simplifying this into a manageable quadratic equation, we have $x^2 + 2bx + b^2 - ab = 0$, where ignoring the negative root leaves us with $x = -b + \sqrt{b^2 - b^2 + ab} = -b + \sqrt{ab}$.

However, $EF = x + b = -b + \sqrt{ab} + b = \sqrt{ab} = a \text{ Ⓖ } b$, which shows that EF represents the geometric mean.

The harmonic mean $a \text{ Ⓗ } b = \frac{2ab}{a+b}$ between the two bases a and b of a trapezoid is much easier to identify. It is simply the line m_H parallel to the bases and containing the point of intersection of the diagonals. (See figure 4.18.) We can show that this point of intersection, S, of the diagonals divides EF into two equal segments so that $ES = u = v = SF = \frac{ab}{a+b} = \frac{m_H}{2}$, as shown in figure 4.19.

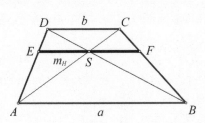

Figure 4.18   Figure 4.19

With all corresponding angles equal because of the parallel bases, we have $\triangle ABS \sim \triangle CDS$, and with $AC = AS + SC = e_1 + e_2$, and $BD = BS + SD = f_1 + f_2$. We then get $\frac{e_1}{e_2} = \frac{f_1}{f_2} = \frac{a}{b}$. (See figure 4.19.)

We also have equal areas in that $A_{\triangle ADS} = A_{\triangle BCS}$. This results from $A_{\triangle ABC} = A_{\triangle ABD} = \frac{1}{2} a \cdot h$ (where $h = h_1 + h_2$) and removing triangle ABS from these two equal areas ($A_{\triangle ABC} = A_{\triangle ABD}$). In other words, $A_{\triangle ADS} = A_{\triangle ABD} - A_{\triangle ABS} = A_{\triangle ABC} - A_{\triangle ABS} = A_{\triangle BCS}$.

From $\triangle ABS \sim \triangle CDS$, we have $\frac{AS}{SC} = \frac{BS}{SD}$, or $\frac{e_1}{e_2} = \frac{f_1}{f_2}$. From $\frac{e_1}{e_2} = \frac{a}{b}$, or $e_2 = e_1 \cdot \frac{b}{a}$.

We can also show that $\frac{FS}{CD} = \frac{AS}{AC} = \frac{AS}{AS+SC}$. Whereupon $\frac{u}{b} = \frac{e_1}{e_1+e_2}$, which we can write as: $u = b \cdot \frac{e_1}{e_1+e_2} = b \cdot \frac{e_1}{e_1 + e_1 \cdot \frac{b}{a}} = b \cdot \frac{e_1}{e_1 \cdot \left(1 + \frac{b}{a}\right)} = \frac{b}{1 + \frac{b}{a}} = \frac{b}{\frac{a}{a} + \frac{b}{a}} = \frac{ab}{a+b}$.

Since $\frac{f_1}{f_2} = \frac{a}{b}$, analogously, we can show that $v = \frac{ab}{a+b}$; therefore, $u = v = \frac{ab}{a+b} = \frac{m_H}{2}$, which is $EF = a \oplus b = \frac{2ab}{a+b}$, the harmonic mean.

We can find the length on the trapezoid $ABCD$ of the root-mean-square $a \circledR b = \sqrt{\frac{a^2+b^2}{2}}$ in terms of the base lengths a and b as the line segment m_R, which partitions the trapezoid into two trapezoids of equal area, namely, $ABFE$ and $EFDC$ as shown in figure 4.20.

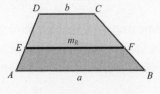

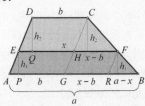

Figure 4.20   Figure 4.21

In figure 4.21, we have $h_1 = EP$, $h_2 = DQ$, and $x = EF$. Since the trapezoids $ABFE$ and $EFCD$ are equal in area, we have $A_{ABFE} = \frac{a+x}{2} \cdot h_1$ and $A_{EFCD} = \frac{x+b}{2} \cdot h_2$.

Thus, $A_{ABFE} = A_{EFCD}$, yields $\frac{a+x}{2} \cdot h_1 = \frac{x+b}{2} \cdot h_2$, or $\frac{h_1}{h_2} = \frac{x+b}{a+x}$. Since $CG \parallel FR \parallel AD$, $b = CD = AG = EH$, $x = EF = AR$, $x - b = FH = GR$, and $a - x = BR$, we then have $\triangle BFR \sim \triangle FCH$, and $\frac{h_1}{h_2} = \frac{BR}{FH} = \frac{a-x}{x-b}$.

Therefore, $\frac{x+b}{a+x} = \frac{a-x}{x-b}$, or simplified, we get, $(a + x)(a - x) = (x + b)(x - b)$, or $a^2 - x^2 = x^2 - b^2$, giving us $a^2 + b^2 = 2x^2$, which then produces our desired result, namely, $x = \sqrt{\frac{a^2+b^2}{2}} = a \circledR b$, which is the root-mean-square.

An alternate approach to arrive at this conclusion about the root-mean-square is shown in figure 4.22, where with the two equal-area trapezoids $ABEF$ and $FECD$ we have $\triangle ABS \sim \triangle FES \sim \triangle DCS$.

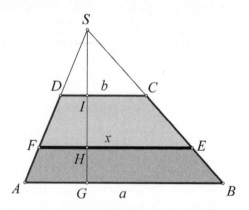

Figure 4.22

We then have $AB : FE : DC = a : x : b$.

The ratio of the areas is then $A_{\triangle ABS} : A_{\triangle FES} : A_{\triangle DCS} = a^2 : x^2 : b^2$.

Since $A_{ABEF} = A_{FECD}$, we have $A_{\triangle ABS} - A_{\triangle FES} = A_{\triangle FES} - A_{\triangle DCS}$, which, from the previous ratio, gives us $a^2 - x^2 = b^2 - x^2$, from which we get $x = \sqrt{\frac{a^2+b^2}{2}} = a \circledR b$, the root-mean-square.

When we place the four means on the same trapezoid that we have just individually considered as the length of various parallel lines to the bases, we get the diagram shown in figure 4.23. Here we have, in sized order (from smallest to largest):

- *EF* is the harmonic mean, m_H (the parallel through the point of intersection of the diagonals)
- *GH* is the geometric mean, m_G (the parallel that divides the trapezoid into two similar trapezoids)
- *JK* is the arithmetic mean, m_A (the parallel that joins the midpoints of the two sides of the trapezoid)
- *MN* is the root-mean-square, m_R (the parallel that divides the trapezoid into two equal-area trapezoids)

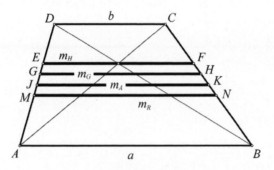

Figure 4.23

There are number of other means that we can present here but that are rarely seen or used, however, they enhance the curiosity of the concept of a mean. First, there is the *contraharmonic mean*, which we will define as $a \, \text{©} \, b = \frac{a^2+b^2}{a+b}$, and it can be seen geometrically in figure 4.24 as:

$$BM + ME = BM + (CM - CE) = (a \, \text{Ⓐ} \, b) + [(a \, \text{Ⓐ} \, b) - (a \, \text{Ⓗ} \, b)] = 2(a \, \text{Ⓐ} \, b) - (a - \text{Ⓗ} \, b)$$
$$= 2 \cdot \frac{a+b}{2} - \frac{2ab}{a+b} = \frac{(a+b)(a+b)}{a+b} - \frac{2ab}{a+b} = \frac{a^2+2ab+b^2-2ab}{a+b} = \frac{a^2+b^2}{a+b} = a \, \text{©} \, b.$$

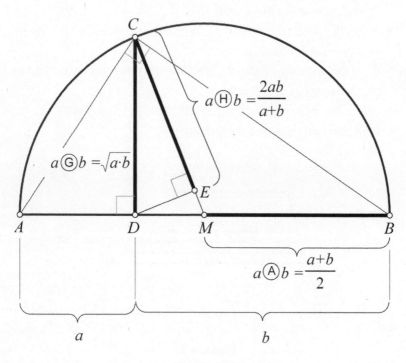

Figure 4.24

We can see this contraharmonic mean in comparison to the other means on the trapezoid in figure 4.25 and represented as $a \; \textcircled{C} \; b = m_C$. There we find that the arithmetic mean, m_A, lies exactly midway between the harmonic mean, m_H, and the contraharmonic mean, m_c. In figure 4.25, we have $K_A K_C = K_A K_H$.

There is also the *Heronian mean*, which we will define as $a \; \textcircled{N} \; b = \frac{a + \sqrt{ab} + b}{3}$. From this we have $a \; \textcircled{N} \; b = \frac{a + \sqrt{ab} + b}{3} = \frac{2}{3} \cdot \frac{a+b}{2} + \frac{1}{3} \cdot \sqrt{ab} = \frac{2}{3} \cdot (a \; \textcircled{A} \; b) + \frac{1}{3} \cdot (a \; \textcircled{G} \; b)$.

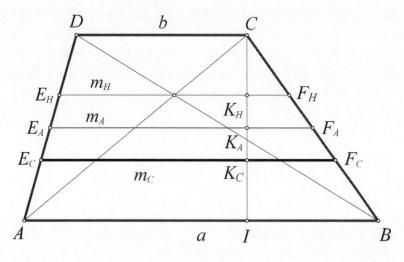

Figure 4.25

When we compare this Heronian mean, m_N, to the arithmetic mean, m_A, and the geometric mean, m_G, we find that it lies between these two means in a ratio of 1:2, so that in figure 4.26 we have $K_A K_N = \frac{1}{3} K_A K_G$.

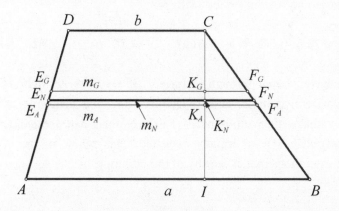

Figure 4.26

Lastly, we will consider the *controidal mean*, which we will denote as $TM(a, b) = a \;ⓉT\; b = \frac{2(a^2 + ab + b^2)}{3(a+b)}$. We can see how this mean, m_T, manifests itself geometrically in figure 4.27, where we have a trapezoid and its center of gravity, point T, which we can locate by taking the intersection of the lines joining the midpoints of the two parallel bases, and the line joining the endpoints of the extension of the bases—extended in opposite directions a distance equal to the opposite bases. In figure 4.27, we have $a = AB$ and $b = CD$, which are used to extend these bases.

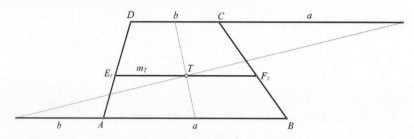

Figure 4.27

In summary, for all positive real numbers a and b, we can make the comparison of all the means as follows:

$$\frac{2ab}{a+b} \leq \sqrt{a \cdot b} \leq \frac{a+\sqrt{ab}+b}{3} \leq \frac{a+b}{2} \leq \frac{2(a^2+ab+b^2)}{3(a+b)} \leq \sqrt{\frac{a^2+b^2}{2}} \leq \frac{a^2+b^2}{a+b}$$

This can be written symbolically as:

$$a \oplus b \leq a \circledcirc b \leq \quad a \circledR b \quad \leq a \circledA b \leq \quad a \circledT b \quad \leq a \circledR b \quad \leq a \copyright b.$$

We have shown through a variety of ways—with right triangles, rectangles, a hyperbola, and trapezoids—how the measures of central tendency (that is, the various means) can be compared in size geometrically, something that is not normally expected. This is one of the neglected curiosities in mathematics worthy of our attention.

Chapter 5

AN UNUSUAL WORLD OF FRACTIONS

A s we begin our journey through the world of fractions—albeit from a somewhat atypical viewpoint, we will consider an often-misunderstood aspect of fractions—how they relate to each other in an unusual way. Later we will see how fractions are used to explore some rather-curious mathematical relationships. We begin by refreshing our "understanding" of fractions.

UNDERSTANDING FRACTIONS

Suppose we have a square piece of paper that we will cut multiple times. The square measures 10 cm by 10 cm, and we first cut off a strip parallel to one side, which is $\frac{1}{6}$ of its area, as shown in figure 5.1 (with the cuts numbered consecutively). We then continue to cut away $\frac{1}{5}$ of the remaining area of the square, then $\frac{1}{4}$ of the remaining area, then $\frac{1}{3}$ of the remaining area, and lastly $\frac{1}{2}$ of that remaining area. We need to determine what part of the area of the original square is left after these five pieces have been cut away. In figure 5.1, we show this as the nonshaded region.

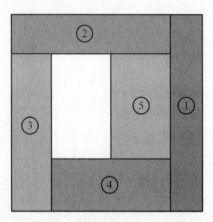

Figure 5.1

Approaching this logically, when we cut away $\frac{1}{6}$ of the square, we were left with numbered $\frac{5}{6}$. When we cut $\frac{1}{5}$ of the area of the remaining piece of paper, we were left with $\frac{4}{5}$ of that remaining piece of paper. However, when figured on the original square piece of paper, we now have left $\frac{4}{5} \cdot \frac{5}{6}$ of the original square piece of paper. When we then cut $\frac{1}{4}$ of the remaining piece of paper, we are left with $\frac{3}{4}$ of that remaining piece of paper, or $\frac{3}{4} \cdot \frac{4}{5} \cdot \frac{5}{6}$ of the original square piece of paper. We now can observe the pattern being followed that the remaining piece (unshaded in figure 5.1) is $1 \cdot \frac{5}{6} \cdot \frac{4}{5} \cdot \frac{3}{4} \cdot \frac{2}{3} \cdot \frac{1}{2} = \frac{1}{6}$ of the area of the original square.

Once the pattern was developed, and the reasoning clear, the original question becomes rather simple. However, in the way it was presented, it may lead us down the wrong path for solution.

Unit fractions are perhaps the basic building blocks for understanding fractions. From ancient times, the unit fractions were the most easy to comprehend, as they represented one piece from a collection or set of given equal pieces. We now consider how these unit fractions—those in the form $\frac{1}{n}$, where n is a natural number greater than zero—from a new viewpoint in mathematics.

THE HARMONIC TRIANGLE—
A ROLE FOR UNIT FRACTIONS

The use of unit fractions goes back to antiquity, where the Egyptians relied on unit fractions almost exclusively for all their measurement—the one exception was $\frac{2}{3}$. For quantities that did not lend themselves to being measured as a unit fraction, the Egyptians merely added several unit fractions to represent that quantity—and, as mentioned earlier, included $\frac{2}{3}$ to sums when needed. We, however, will be using unit fractions in a rather-unusual fashion. We are going to create a triangular arrangement of unit fractions, one that was first discovered by the German mathematician Gottfried Wilhelm Leibniz (1646–1716), who, by the way, was also responsible for having redeveloped calculus for modern times, including the calculus nomenclature we use today. We say "redeveloped" since we now know that Eudoxus (408–355 BCE) had a process of exhaustion that anticipated calculus, yet Isaac Newton (1642–1727) and Leibniz developed it independently without having any knowledge of Eudoxus's work.

Let us first recall that a unit fraction is a fraction in the form $\frac{1}{n}$, where n is any positive natural number. We are now going to set up a triangular arrangement of unit fractions in a way so that the outer obliques will form a harmonic sequence[1] and each member of the triangular arrangement of unit fractions will be such that the sum of the two fractions below it—one to the right and the one to the left of the member—will be equal to that fraction. We will call this a *harmonic triangle*. For example, see how these fractions are placed in figure 5.2:

$$\frac{1}{2} = \frac{1}{3} + \frac{1}{6} = \frac{1}{6} + \frac{1}{3}.$$

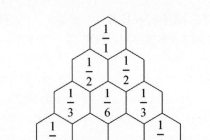

Figure 5.2

In more general terms, we show in figure 5.3 how the sum of the fractions $\frac{1}{x}$ and $\frac{1}{y}$, are placed, which results in $\frac{1}{z} = \frac{1}{x} + \frac{1}{y}$.

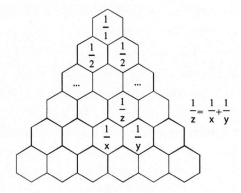

Figure 5.3

In figure 5.4, we show where the name *harmonic* comes from by inspecting the outer obliques and noticing their sum. This is a harmonic series, which we have seen earlier.

$$1 + \frac{1}{2} + \frac{1}{3} + \frac{1}{4} + \frac{1}{5} + \frac{1}{6} + \cdots$$

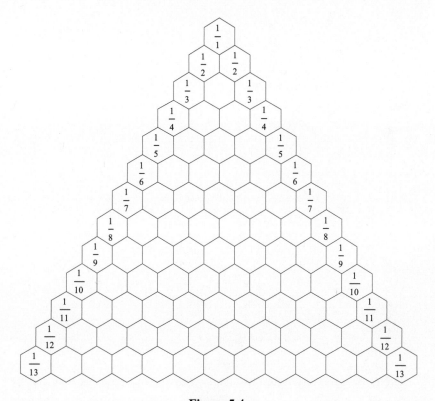

Figure 5.4

In figure 5.5, we provide a thirteen-row harmonic triangle, which should provide a better understanding of the harmonic triangle.

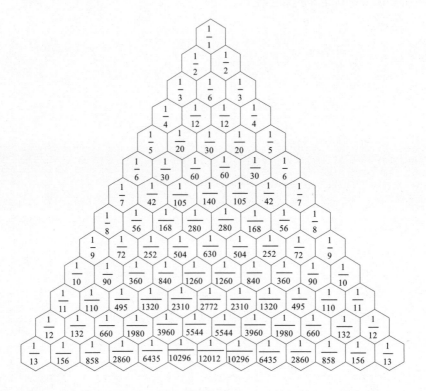

Figure 5.5

There are probably countless arrangements and patterns that can be discovered in the harmonic triangle. As we mentioned earlier, to find a new member of the harmonic triangle, we would set up the following equation: $\frac{1}{3} = \frac{1}{4} + \frac{1}{x}$. Then solve for $\frac{1}{x}$, $\frac{1}{x} = \frac{1}{3} - \frac{1}{4} = \frac{4-3}{3 \cdot 4} = \frac{1}{12}$, as shown in figure 5.6.

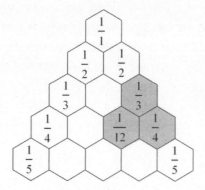

Figure 5.6

In more general terms, we would have $\frac{1}{n} = \frac{1}{n+1} + \frac{1}{n(n+1)}$, and thus, $\frac{1}{n(n+1)} = \boxed{\frac{1}{n} \cdot \frac{1}{n+1}} = \boxed{\frac{1}{n} - \frac{1}{n+1}}$.

Or put another way, the difference between consecutive unit fractions is equal to their product. (See figure 5.7.) This, in and of itself, is a long-neglected relationship that seems to have escaped many during their school years, when working with fractions was more of an exercise of drill rather than an appreciation of their various interrelationships.

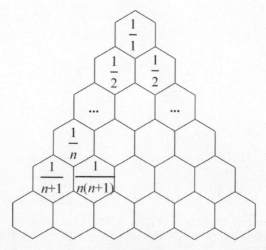

Figure 5.7

Let us continue our journey along our investigation of the harmonic triangle, which is an arrangement of unit fractions that gives us some new appreciation of unit fractions. For example, we can see a further interesting pattern on the harmonic triangle, if we look at the second oblique in figure 5.8, where we notice the sequence (a_n) with $(a_n) = \frac{1}{n(n+1)}$, whose members can be represented as:

$$\frac{1}{2} = \frac{1}{1 \cdot 2},$$

$$\frac{1}{6} = \frac{1}{2 \cdot 3},$$

$$\frac{1}{12} = \frac{1}{3 \cdot 4},$$

$$\frac{1}{20} = \frac{1}{4 \cdot 5},$$

$$\frac{1}{30} = \frac{1}{5 \cdot 6}, \ldots$$

This is a curious relationship that will help us understand other relationships as we progress along our investigation of the harmonic triangle.

Figure 5.8

We notice from the harmonic triangle shown in figure 5.8 that $\frac{1}{30} = \frac{1}{60} + \frac{1}{60}$. However, if we go down one row, we see the fraction $\frac{1}{30}$ once again appear, but this time $\frac{1}{30} = \frac{1}{42} + \frac{1}{105}$. To justify this relationship is a bit complex, but we shall provide it here for the ambitious reader.

We refer back to figure 5.8 and notice that the first fraction in the first row is $a_{1,1} = \frac{1}{1}$, and the first fraction in the nth row is $a_{n,1} = \frac{1}{n}$. Taking this further, for example, the second fraction in the fourth row is $a_{4,2} = \frac{1}{12}$. (See figure 5.9.)

From our established relationship $\frac{1}{x} + \frac{1}{y} = \frac{1}{z}$, we can also write the recursive equation: $a_{n, k-1} + a_{n, k} = a_{n-1, k-1}$, or, written another way, $a_{n,k} = a_{n-1, k-1} - a_{n, k-1}$.

$a_{n,k} \diagdown k$	1	2	3	4	5	6	7	8
1	$\frac{1}{1}$	—	—	—	—	—	—	—
2	$\frac{1}{2}$	$\frac{1}{2}$	—	—	—	—	—	—
3	$\frac{1}{3}$	$\frac{1}{6}$	$\frac{1}{3}$	—	—	—	—	—
4	$\frac{1}{4}$	$\frac{1}{12}$	$\frac{1}{12}$	$\frac{1}{4}$	—	—	—	—
5	$\frac{1}{5}$	$\frac{1}{20}$	$\frac{1}{30}$	$\frac{1}{20}$	$\frac{1}{5}$	—	—	—
6	$\frac{1}{6}$	$\frac{1}{30}$	$\frac{1}{60}$	$\frac{1}{60}$	$\frac{1}{30}$	$\frac{1}{6}$	—	—
7	$\frac{1}{7}$	$\frac{1}{42}$	$\frac{1}{105}$	$\frac{1}{140}$	$\frac{1}{105}$	$\frac{1}{42}$	$\frac{1}{7}$	—
8	$\frac{1}{8}$	$\frac{1}{56}$	$\frac{1}{168}$	$\frac{1}{280}$	$\frac{1}{280}$	$\frac{1}{168}$	$\frac{1}{56}$	$\frac{1}{8}$

Figure 5.9

Representing the term $a_{n,k}$ is rather complex, but we provide it here for completion of this explanation.

$$a_{n,k} = \frac{1}{k \cdot \binom{n}{k}} = \frac{1}{n \cdot \binom{n-1}{k-1}}, \text{ where } \binom{n}{k} \text{ is the binomial coefficient } (n, k \in \mathbf{N}).$$

Therefore, for $n = 7$, $k = 2$, we get the following: $a_{7,2} + a_{7,3} = a_{6,2}$, which gives us the desired $\frac{1}{42} + \frac{1}{105} = \frac{1}{30}$.

Generalizing this from our earlier discoveries, we get the relationship:

$$\frac{1}{n} = \frac{1}{n+1} + \frac{1}{n(n+1)}$$

We can also look at the sum of each row to see if there is a pattern that evolves. The following table shows the sum of the first six rows.

Row	1	2	3	4	5	6
Sum	$1 = \frac{60}{60}$	$1 = \frac{60}{60}$	$\frac{5}{6} = \frac{50}{60}$	$\frac{2}{3} = \frac{40}{60}$	$\frac{8}{15} = \frac{32}{60}$	$\frac{13}{30} = \frac{26}{60}$

In figure 5.10, we highlight the sixth row and consider its sum as follows:

$$\frac{1}{6} + \frac{1}{30} + \frac{1}{60} + \frac{1}{60} + \frac{1}{30} + \frac{1}{6} = \frac{13}{30} = \frac{26}{60} = 0.4333333333\ldots$$

Now going to the thirteenth row, we find the sum of the elements in that row to be as follows:

$$\frac{1}{13} + \frac{1}{156} + \frac{1}{858} + \frac{1}{2860} + \ldots + \frac{1}{2860} + \frac{1}{858} + \frac{1}{156} + \frac{1}{13} = \frac{15341}{90090} = 0.1702852702\ldots$$

We can deduce from this that as we move down the rows, the sum of the elements in each of the rows gets smaller as the number of unit fractions increases.

Now let us inspect the sum of the elements in the various obliques. We already established that the sum of the elements of the first oblique, a harmonic sequence, $1 + \frac{1}{2} + \frac{1}{3} + \frac{1}{4} + \frac{1}{5} + \frac{1}{6} + \ldots$, continuously increases because we are adding positive values to the sum (that is, it diverges), and therefore it approaches infinity. We can write this as $\sum_{k=1}^{\infty}\frac{1}{k}=\infty$.

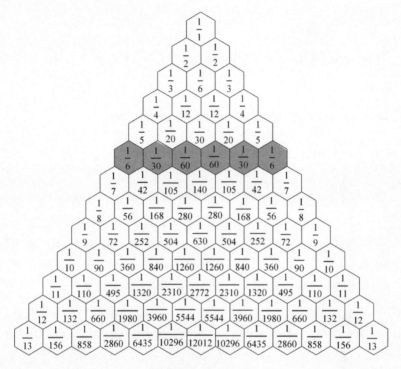

Figure 5.10

Once again we focus on the second oblique,

$$\frac{1}{2}+\frac{1}{6}+\frac{1}{12}+\frac{1}{20}+\frac{1}{30}+\frac{1}{42}+\ldots,$$

which we highlighted in figure 5.8. We will have to take a moment to consider how best to find this sum. As we look at each of the members of this oblique, we recall the following pattern emerging:

$$\frac{1}{2}=\frac{1}{1\cdot2},$$

$$\frac{1}{6}=\frac{1}{2\cdot3},$$

$$\frac{1}{12} = \frac{1}{3 \cdot 4},$$

$$\frac{1}{20} = \frac{1}{4 \cdot 5}, \ldots$$

Therefore, for the nth member of this second oblique, we have the term

$$\frac{1}{n \cdot (n+1)}.$$

If we consider the first four terms, we get the sum

$$\frac{1}{2} + \frac{1}{6} + \frac{1}{12} + \frac{1}{20} = \frac{4}{5} = 0.8.$$

The first twelve terms of this second oblique gives us the following sum:

$$\frac{1}{2} + \frac{1}{6} + \frac{1}{12} + \ldots + \frac{1}{132} + \frac{1}{156} = \frac{12}{13} \approx 0.923.$$ (It appears as though this sum approaches 1.)

From this we can set up the general n^{th} term:

$$\frac{1}{2} + \frac{1}{6} + \frac{1}{12} + \frac{1}{20} + \ldots + \frac{1}{n \cdot (n+1)} = \sum_{k=1}^{n} \frac{1}{k(k+1)} = \frac{n}{n+1}.$$

Continuing this, we will notice that the more terms we add to our sum, the closer we get to 1. If we carry this to infinity, we would get the following:

$$\sum_{k=1}^{\infty} \frac{1}{k(k+1)}.$$

Moving on to the third oblique, we consider the sum of the elements in that oblique.

$$\frac{1}{3} + \frac{1}{12} + \frac{1}{30} + \frac{1}{60} + \frac{1}{105} + \frac{1}{168} + \ldots$$

Once again, we need to look at the individual terms in order to be able to establish a pattern for adding these terms. Consider each of the terms as follows:

$$\frac{1}{3} = \frac{2}{1 \cdot 2 \cdot 3},$$

$$\frac{1}{12} = \frac{2}{2 \cdot 3 \cdot 4},$$

$$\frac{1}{30} = \frac{2}{3 \cdot 4 \cdot 5},$$

$$\frac{1}{60} = \frac{2}{4 \cdot 5 \cdot 6}, \ldots$$

The pattern leads us to generalizing this in the following fashion:

$$\frac{2}{n \cdot (n+1) \cdot (n+2)}.$$

Again looking at the first four terms, we get the following sum:

$$\frac{1}{3} + \frac{1}{12} + \frac{1}{30} + \frac{1}{60} = \frac{7}{15} \approx 0.467.$$

Taking the sum of the first eleven terms gives us the following sum:

$$\frac{1}{3} + \frac{1}{12} + \frac{1}{30} + \cdots + \frac{1}{660} + \frac{1}{858} = \frac{77}{156} \approx 0.494.$$

Proceeding to get the sum of n terms, we would have the following:

$$\frac{1}{3} + \frac{1}{12} + \frac{1}{30} + \frac{1}{60} + \ldots + \frac{2}{n \cdot (n+1) \cdot (n+2)} = \sum_{k=1}^{n} \frac{2}{k(k+1)(k+2)} = \frac{1}{2} - \frac{1}{(n+1) \cdot (n+2)}.$$

The more terms we take, the closer we will get to a sum of $\frac{1}{2}$. Therefore, when we take the sum of an infinite number of terms of the third oblique, the sum would be: $\sum_{k=1}^{\infty} \frac{2}{k(k+1)(k+2)} = \frac{1}{2}$. The fourth oblique is even more complicated than the previous ones. The general term for the nth member is $\frac{6}{n \cdot (n+1) \cdot (n+2) \cdot (n+3)}.$

The sum of the first ten terms of this oblique is:

$$\frac{1}{4} + \frac{1}{20} + \frac{1}{60} + \ldots + \frac{1}{1980} + \frac{1}{2860} = \frac{95}{286} \approx 0.332.$$

We can then see that as this series increases, it approaches $\frac{1}{3}$.

Then the sum of the first n unit fractions of the fourth oblique is as follows:

$$\sum_{k=1}^{n} \frac{6}{k(k+1) \cdot (k+2) \cdot (k+3)} = \frac{1}{3} - \frac{2}{(n+1) \cdot (n+2) \cdot (n+3)}$$

We can see that the infinite sum of this series is: $\sum_{k=1}^{\infty} \frac{6}{k(k+1) \cdot (k+2) \cdot (k+3)} = \frac{1}{3}.$

By now you might have noticed that each term in the harmonic sequence of the outer oblique is the sum of terms that begins immediately below it and continues downward along the oblique, as for example, $\frac{1}{6}=\frac{1}{7}+\frac{1}{56}+\frac{1}{252}+\frac{1}{840}+\frac{1}{2310}+\frac{1}{5544}+\frac{1}{10296}+\ldots$

At this point, having seen several unexpected relationships and patterns on the harmonic triangle, we now approach a much more familiar triangle arrangement of numbers, known as the *Pascal triangle*, named for Blaise Pascal (1623–1662), who, in 1653, invented his arithmetical triangle, although it was not published until after his death. This triangular arrangement of numbers was first described by the Arabian mathematician Omar Khayyam (1048–1122), but it did not appear in print until 1303, in *The Valuable Mirror of the Four Elements* by the Chinese mathematician Chu Shih-Chieh (ca. 1270–1330). Yet, in the Western world, Pascal is credited with having discovered it without reference to any previous documents. We present this Pascal triangle in figure 5.11.

$$
\begin{array}{c}
1 \quad = 2^0 \\
1 \quad 1 \quad = 2^1 \\
1 \quad 2 \quad 1 \quad = 2^2 \\
1 \quad 3 \quad 3 \quad 1 \quad = 2^3 \\
1 \quad 4 \quad 6 \quad 4 \quad 1 \quad = 2^4 \\
1 \quad 5 \quad 10 \quad 10 \quad 5 \quad 1 \quad = 2^5 \\
1 \quad 6 \quad 15 \quad 20 \quad 15 \quad 6 \quad 1 \quad = 2^6 \\
1 \quad 7 \quad 21 \quad 35 \quad 35 \quad 21 \quad 7 \quad 1 \quad = 2^7 \\
1 \quad 8 \quad 28 \quad 56 \quad 70 \quad 56 \quad 28 \quad 8 \quad 1 \quad = 2^8 \\
1 \quad 9 \quad 36 \quad 84 \quad 126 \quad 126 \quad 84 \quad 36 \quad 9 \quad 1 \quad = 2^9 \\
1 \quad 10 \quad 45 \quad 120 \quad 210 \quad 252 \quad 210 \quad 120 \quad 45 \quad 10 \quad 1 \quad = 2^{10}
\end{array}
$$

Figure 5.11

The Pascal triangle is very rich in the number of patterns and relationships that it exhibits. Perhaps one of its most useful applications is providing the coefficients for binomial expansion. However, in viewing the triangular arrangement, the first five rows visually give us the first five powers of 11. The sum of each row gives us increasingly powers of 2, as seen in figure 5.11. Using "awkward" obliques, as shown in figure 5.12

below, and taking the sum of those numbers, we get the *Fibonacci numbers* (1, 1, 2, 3, 5, 8, 13, 21, 34, 55, 89, . . .) in sequence on the triangle. This is merely scratching the surface of the many relationships that can be found on the Pascal triangle.

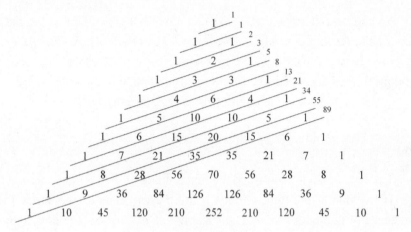

Figure 5.12

Let us now investigate how these two famous triangular arrangements of numbers, the Pascal triangle and the harmonic triangle, can be related. We begin with one relationship where we will compare the sixth row of each of these triangles and establish a truly unexpected relationship.

The sixth row of the Pascal triangle reads 1, 5, 10, 10, 5, 1. (See figure 5.11). Now let us look at the first member of the sixth row of the harmonic triangle, $\frac{1}{6}$. We will now divide this number, $\frac{1}{6}$, by each of the numbers in the sixth row of the Pascal triangle as follows:

$$\frac{1}{6} \div 1 = \frac{1}{6},$$

$$\frac{1}{6} \div 5 = \frac{1}{30},$$

$$\frac{1}{6} \div 10 = \frac{1}{60},$$

$$\frac{1}{6} \div 10 = \frac{1}{60},$$

$$\frac{1}{6} \div 5 = \frac{1}{30},$$

$$\frac{1}{6} \div 1 = \frac{1}{6}.$$

Lo and behold (amazingly!), we end up with the members of the sixth row of the harmonic triangle. There are other relationships that we can discover between these two famous triangles. For example, the second oblique of the Pascal triangle is the sequence of natural numbers: 1, 2, 3, 4, 5, . . . When we look at the reciprocals of each of these numbers,

$$1, \frac{1}{2}, \frac{1}{3}, \frac{1}{4}, \frac{1}{5}, \cdots,$$

we find that we have established the first oblique of the harmonic triangle.

If we look at the third oblique of the Pascal triangle, we notice the sequence of numbers known as the *triangular numbers*, since dots representing these numbers,

$$1, 3, 6, 10, 15, \ldots, \frac{n(n+1)}{2}, \ldots,$$

can be placed into the form of an equilateral triangle, as shown in figure 5.13.

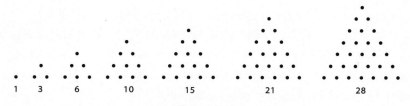

Figure 5.13

We will now take the reciprocals of each of these numbers to get $1, \frac{1}{3}, \frac{1}{6}, \frac{1}{10}, \frac{1}{15}, \ldots$ Then we take one-half of each of these numbers, and we get the following sequence:

$$\frac{1}{2}, \frac{1}{6}, \frac{1}{12}, \frac{1}{20}, \frac{1}{30}, \ldots, \frac{1}{n \cdot (n+1)}, \ldots,$$ which we have shown previously in figure 5.8.

When we take the sum of the reciprocals of the triangular numbers, we get:

$$1 + \frac{1}{3} + \frac{1}{6} + \frac{1}{10} + \ldots + \frac{2}{n(n+1)} + \ldots = 2$$

The sum of the numbers of the harmonic triangle's second oblique is:

$$\frac{1}{2} + \frac{1}{6} + \frac{1}{12} + \frac{1}{20} + \ldots + \frac{1}{n(n+1)} + \ldots = 1.$$

The number 1 is just above the first member of this series $\frac{1}{2}$, and if we were to begin with this, we would have equal sums.

In the fourth oblique of the Pascal triangle, we find the *tetrahedral numbers*[2]:

$$1, 4, 10, 20, 35, 56, \ldots, \frac{n(n+1)(n+2)}{6}, \ldots$$

If we now take the reciprocals of each of these numbers and multiply them by $\frac{1}{3}$, which is the first number of the third oblique of the harmonic triangle, we get the following sequence:

$$\frac{1}{3}, \frac{1}{12}, \frac{1}{30}, \frac{1}{60}, \frac{1}{105}, \ldots, \frac{2}{n \cdot (n+1) \cdot (n+2)}, \ldots$$

We also notice a relationship between the sum of the reciprocals of the tetrahedral numbers,

$$1 + \frac{1}{4} + \frac{1}{10} + \frac{1}{20} + \ldots + \frac{6}{n(n+1)(n+2)} + \ldots = \frac{3}{2},$$

and the sum of the third oblique of the harmonic triangle (which we have already seen earlier):

$$\frac{1}{3} + \frac{1}{12} + \frac{1}{30} + \frac{1}{60} + \ldots + \frac{2}{n \cdot (n+1) \cdot (n+2)} + \ldots = \frac{1}{2}.$$

Curiously, they differ by 1.

Taking this one step further to the fifth oblique of the Pascal triangle, we find the *pentatope numbers*[3]:

$$1, 5, 15, 35, 70, \ldots, \frac{n(n+1)(n+2)(n+3)}{24}, \ldots$$

We can see that the fourth oblique of the harmonic triangle gives us the following:

$$\frac{1}{4} + \frac{1}{20} + \frac{1}{60} + \frac{1}{140} + \ldots + \frac{6}{n \cdot (n+1) \cdot (n+2) \cdot (n+3)} + \ldots = \sum_{n=1}^{\infty} \frac{6}{n(n+1) \cdot (n+2) \cdot (n+3)} = \frac{1}{3}.$$

There are numerous curious relationships that can be found between the obliques of these two famous triangles: the Pascal triangle and the harmonic triangle. Figure 5.14 shows a comparison of the general terms of each.

n	Pascal Triangle with the General Term a_n	Harmonic Triangle with the General Term a_n
1ˢᵗ Oblique	1	$\dfrac{1}{n}$
2ⁿᵈ Oblique	n	$\dfrac{1}{n(n+1)}$
3ʳᵈ Oblique	$\dfrac{n(n+1)}{2}$	$\dfrac{2}{n(n+1)(n+2)}$
4ᵗʰ Oblique	$\dfrac{n(n+1)(n+2)}{6}$	$\dfrac{6}{n(n+1)(n+2)(n+3)}$
5ᵗʰ Oblique	$\dfrac{n(n+1)(n+2)(n+3)}{24}$	$\dfrac{24}{n(n+1)(n+2)(n+3)(n+4)}$
. . .	. . .	. . .
k^{th} Oblique ($k > 1$)	$\dfrac{n(n+1)(n+2) \cdot \ldots \cdot (n+k-2)}{(k-1)!}$	$\dfrac{(k-1)!}{n(n+1)(n+2) \cdot \ldots \cdot (n+k-1)}$

Figure 5.14

Finally, we look at the limiting value of the infinite series for the kth oblique of the harmonic triangle, which happens to be $\dfrac{1}{k-1}$ and is also the first member of the previous row (or oblique).

A summary of these limiting values is shown in figure 5.15.

n	Harmonic Triangle	
	General term a_n	Limiting Value
1st Oblique	$\dfrac{1}{n}$	∞
2nd Oblique	$\dfrac{1}{n(n+1)}$	1
3rd Oblique	$\dfrac{2}{n(n+1)(n+2)}$	$\dfrac{1}{2}$
4th Oblique	$\dfrac{6}{n(n+1)(n+2)(n+3)}$	$\dfrac{1}{3}$
5th Oblique	$\dfrac{24}{n(n+1)(n+2)(n+3)(n+4)}$	$\dfrac{1}{4}$
...	...	...
kth Oblique ($k > 1$)	$\dfrac{(k-1)!}{n(n+1)(n+2)\cdot\ldots\cdot(n+k-1)}$	$\dfrac{1}{k-1}$

Figure 5.15

A VISIT TO AN UNUSUAL WORLD OF FRACTIONS

When we think of addition and multiplication of fractions, perhaps we recall the multiplication process more readily than we do the addition process. This is largely because the multiplication process requires us merely to multiply the two numerators and the two denominators to get the product of the two fractions, as you can see in the example below:

$$\frac{1}{2}\cdot\frac{1}{3}=\frac{1\cdot1}{2\cdot3}=\frac{1}{6}.$$

However, you may recall that in order to add two fractions, one needs to find a common denominator; then the numerators can be added and placed over this common denominator. Consider what might result if we were to add two fractions using the algorithm for multiplication. The result would look like this:

$$\frac{1}{2}+\frac{1}{3}=\frac{1+1}{2+3}=\frac{2}{5}.$$

We can clearly see that this process is incorrect, since the sum of these two fractions is less than one of the fractions. In other words, we get a meaningless result. Were we to use this process, we would encounter other difficulties. For example, if, in such an "addition," we would replace one of the fractions with its equivalent, we would get a different result with this strange algorithm. Here are two examples of the same "addition" problem, where each of the fractions was changed to an equivalent fraction, and we have two different sums.

$$\frac{1}{2}+\frac{1}{3} = \frac{2}{4}+\frac{1}{3} = \frac{2+1}{4+3}=\frac{3}{7}, \text{ and } \frac{1}{2}+\frac{1}{3} = \frac{1}{2}+\frac{2}{6} = \frac{1+2}{2+6}=\frac{3}{8}.$$

This should thoroughly convince you that this strange algorithm for "addition" is incorrect.

Now, you may ask why would we be demonstrating an incorrect algorithm? Actually, this algorithm will lead us to some very interesting fraction relationships.

In the real world, we can use a soccer game as an example. Supposing a soccer team has a game on Saturday and Sunday, and the lead player scores a goal on Saturday when the team, in total, only had two goals. On Sunday, the same player scored one goal again, while the team, in total, had three goals. We could summarize this player's production as: one out of two goals in the first game and one out of three goals in the second game, or a total of two goals out of the five goals the team scored over the weekend. If we represent this numerically, we get the following: $\frac{1}{2}+\frac{1}{3}=\frac{1+1}{2+3}=\frac{2}{5}$. This should look familiar, as it is the same kind of "addition" algorithm that we encountered above. However, in this context, it has become somewhat meaningful. As a result of this more meaningful manifestation, we will give this kind of fraction—formed by having added the numerators and the denominators—a proper name: the *mediant*.[4]

That is, $\frac{a+c}{b+d}$ is the mediant between the fractions $\frac{a}{b}$ and $\frac{c}{d}$, where a, b, c, and d are natural numbers, with $b \neq 0$ and $d \neq 0$. This strange form of addition, $\frac{a}{b} \oplus \frac{c}{d} = \frac{a+c}{b+d}$, we will designate by the symbol $\oplus$, and is sometimes referred to as *Chuquet addition*, named after the French mathematician Nicolas Chuquet (1445–1488), who may be more famous for having introduced the words *million, billion, trillion,* and so on—often referred to as the *Chuquet system*.

In figure 5.16, we show a list of sixteen mediants resulting from $\frac{a}{b} \oplus \frac{c}{d}$, using only the natural numbers 1 and 2.

a	b	c	d	$\dfrac{a+c}{b+d}$	a	b	c	d	$\dfrac{a+c}{b+d}$
1	1	1	1	1	1	1	1	2	$\dfrac{2}{3}$
2	1	1	1	$\dfrac{3}{2}$	2	1	1	2	1
1	2	1	1	$\dfrac{2}{3}$	1	2	1	2	$\dfrac{1}{2}$
2	2	1	1	1	2	2	1	2	$\dfrac{3}{4}$
1	1	2	1	$\dfrac{3}{2}$	1	1	2	2	1
2	1	2	1	2	2	1	2	2	$\dfrac{4}{3}$
1	2	2	1	1	1	2	2	2	$\dfrac{3}{4}$
2	2	2	1	$\dfrac{4}{3}$	2	2	2	2	1

Figure 5.16

This chart has then generated the following seven fractions

$$\frac{1}{2}, \frac{2}{3}, \frac{3}{4}, \frac{1}{1}, \frac{4}{3}, \frac{3}{2}, \text{ and } \frac{2}{1}.$$

The difference between *adjacent* fractions $\frac{a}{b}$ and $\frac{c}{d}$ is $\frac{ad-bc}{bd}$ and is in lowest terms when $ad - bc = \pm1$. We can call two such fractions *neighbor fractions*.

For example, $\frac{1}{2} - \frac{1}{3} = \frac{1}{6}$, where $ad - bc = 1 \cdot 3 - 2 \cdot 1 = 1$, we can say that $\frac{1}{2}$ and $\frac{1}{3}$ are neighbor fractions, as would be the case for $\frac{1}{3}$ and $\frac{2}{5}$ but not for $\frac{3}{4}$ and $\frac{1}{2}$.

Let's consider some interesting theorems that can be established from this new landscape of fractions.

Theorem 1

When $\frac{a}{b} < \frac{c}{d}$, it will always be true that $\frac{a}{b} < \frac{a+c}{b+d} < \frac{c}{d}$, where a, b, c, and d are any natural numbers and $c \neq 0$ and $d \neq 0$.

This theorem has resulted from a rather-cute rendition of a classroom experience reported in the *Mathematics Teacher* journal by Laurence Sherzer in 1973. One of his students came up with a rather-novel discovery, and Sherzer, after allowing the student to share his finding with this class (and then verifying the result), decided that this relationship should be called the *McKay theorem*, named for this high-school student, Thomas McKay, with whom Sherzer credited with the discovery.[5]

The proof of this theorem is easily done as follows:

$$\frac{a}{b} < \frac{c}{d} \Rightarrow a \cdot d < b \cdot c \Rightarrow a \cdot d \underline{+ a \cdot b} < b \cdot c \underline{+ a \cdot b} \Rightarrow (b+d) \cdot a < (a+c) \cdot b$$

$$\Rightarrow \frac{a}{b} < \frac{a+c}{b+d}, \frac{a}{b} < \frac{c}{d} \Rightarrow a \cdot d < b \cdot c \Rightarrow a \cdot d \underline{+ c \cdot d} < b \cdot c \underline{+ c \cdot d} \Rightarrow$$

$$(a+c) \cdot d < (b+d) \cdot c \Rightarrow \frac{a+c}{b+d} < \frac{c}{d}.$$

There is also a rather-clever way in which we can show this relationship geometrically. By observing the increased slope of the hypotenuse of the three right triangles shown in figure 5.17, we can see how these fractions compare. This can also be seen by comparing the tangent ratio of the three angles α, β, and γ, as follows: $\tan \alpha = \frac{a}{b} < \tan \gamma = \frac{a+c}{b+d} < \tan \beta = \frac{c}{d}$, and we can visually notice that $\alpha < \gamma < \beta$, thereby further supporting this fractional relationship.

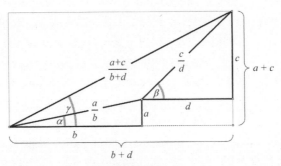

Figure 5.17

Theorem 2

If we have two neighbor fractions $\frac{a}{b}$ and $\frac{c}{d}$, where $\frac{ad - bc}{bd}$ is the difference and is in lowest terms when $ad - bc = \pm 1$, then the mediant fraction, $\frac{a+c}{b+d}$, is a neighbor to each of them.

This can be shown very simply in the following way. The two fractions $\frac{a}{b}$ and $\frac{a+c}{b+d}$ are neighbor fractions if the difference of their cross-product $bc - ad = \pm 1$. So let us now get the analogous product for these two fractions:

$$(b + d) \cdot a - (a + c) \cdot b = ad - bc = -(bc - ad) = \pm 1.$$

Similarly, we will carry out this same process for the two fractions $\frac{a+c}{b+d}$ and $\frac{c}{d}$ as follows:

$$d \cdot (a + c) - c \cdot (b + d) = ad - bc = -(bc - ad) = \pm 1.$$

Therefore, we can see that the mediant fraction is a neighbor to each of the original neighbor fractions.

UNUSUAL FRACTION SEQUENCE—FAREY SEQUENCES

While on the topic of the mediant fraction and its neighbor fractions, we would be remiss not to consider the *Farey sequence*,[6] which is an ordered sequence of fractions beginning with $\frac{0}{1}$ and ending with $\frac{1}{1}$, including all the fractions in that interval, which are in lowest terms. When they are arranged in ascending order, and where no denominator exceeds the value of n, we have a Farey sequence of order n (which we denote with $\mathfrak{F}_n$).

The Farey sequence is named for the British Geologist John Farey Sr. (1766–1826), who wrote about these sequences in the *Philosophical Magazine* in 1816, where he conjectured that each term in the sequence is the mediant of its neighbor fraction. This conjecture was later proved correct by the French mathematician Augustin-Louis Cauchy (1789–1857). Actually, it was later determined that this sequence had previously been known to the French mathematician Charles Haros[7] in 1802, but as is often the case with mathematical advances, the concept was incorrectly attributed to Farey as the originator.

Here we provide the Farey sequences of the first few orders:

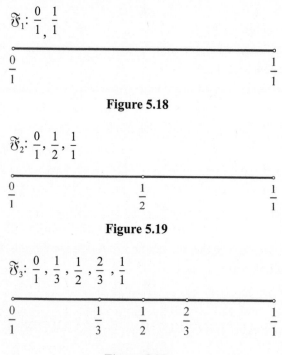

$\mathfrak{F}_1$: $\frac{0}{1}$, $\frac{1}{1}$

$\frac{0}{1}$ $\frac{1}{1}$

Figure 5.18

$\mathfrak{F}_2$: $\frac{0}{1}$, $\frac{1}{2}$, $\frac{1}{1}$

$\frac{0}{1}$ $\frac{1}{2}$ $\frac{1}{1}$

Figure 5.19

$\mathfrak{F}_3$: $\frac{0}{1}$, $\frac{1}{3}$, $\frac{1}{2}$, $\frac{2}{3}$, $\frac{1}{1}$

$\frac{0}{1}$ $\frac{1}{3}$ $\frac{1}{2}$ $\frac{2}{3}$ $\frac{1}{1}$

Figure 5.20

$\mathfrak{F}_4$: $\dfrac{0}{1}, \dfrac{1}{4}, \dfrac{1}{3}, \dfrac{1}{2}, \dfrac{2}{3}, \dfrac{3}{4}, \dfrac{1}{1}$

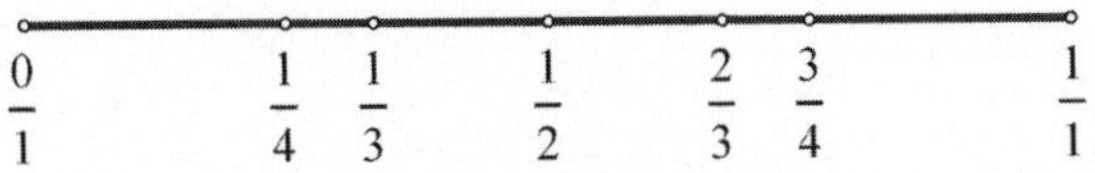

Figure 5.21

$\mathfrak{F}_5$: $\dfrac{0}{1}, \dfrac{1}{5}, \dfrac{1}{4}, \dfrac{1}{3}, \dfrac{2}{5}, \dfrac{1}{2}, \dfrac{3}{5}, \dfrac{2}{3}, \dfrac{3}{4}, \dfrac{4}{5}, \dfrac{1}{1}$

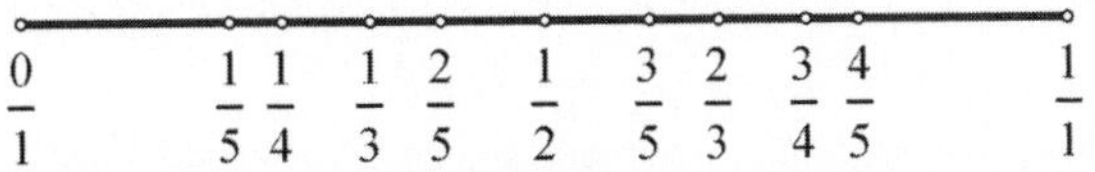

Figure 5.22

$\mathfrak{F}_6$: $\dfrac{0}{1}, \dfrac{1}{6}, \dfrac{1}{5}, \dfrac{1}{4}, \dfrac{1}{3}, \dfrac{2}{5}, \dfrac{1}{2}, \dfrac{3}{5}, \dfrac{2}{3}, \dfrac{3}{4}, \dfrac{4}{5}, \dfrac{5}{6}, \dfrac{1}{1}$

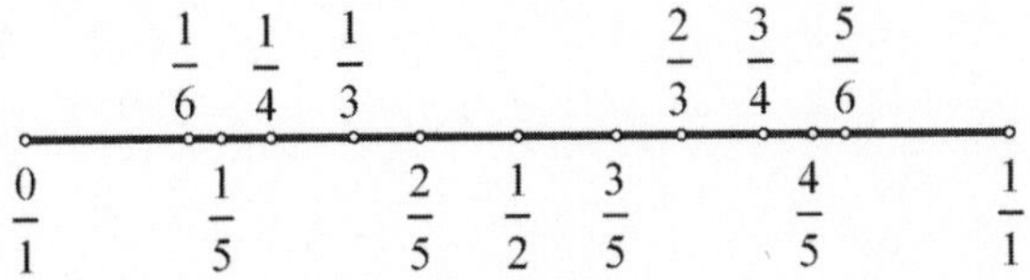

Figure 5.23

$\mathfrak{F}_7$: $\dfrac{0}{1}, \dfrac{1}{7}, \dfrac{1}{6}, \dfrac{1}{5}, \dfrac{1}{4}, \dfrac{2}{7}, \dfrac{1}{3}, \dfrac{2}{5}, \dfrac{3}{7}, \dfrac{1}{2}, \dfrac{4}{7}, \dfrac{3}{5}, \dfrac{2}{3}, \dfrac{5}{7}, \dfrac{3}{4}, \dfrac{4}{5}, \dfrac{5}{6}, \dfrac{6}{7}, \dfrac{1}{1}$

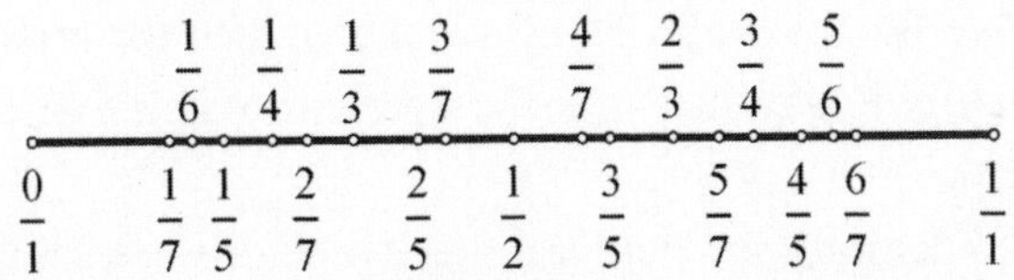

Figure 5.24

We can also admire the set of Farey sequences from $\mathfrak{F}_1$ to $\mathfrak{F}_8$ as shown in figure 5.25.

$\mathfrak{F}_1$: $\frac{0}{1}$ $\quad$ $\frac{1}{1}$

$\mathfrak{F}_2$: $\frac{0}{1}$ $\quad$ $\frac{1}{2}$ $\quad$ $\frac{1}{1}$

$\mathfrak{F}_3$: $\frac{0}{1}$ $\quad$ $\frac{1}{3}$ $\quad$ $\frac{1}{2}$ $\quad$ $\frac{2}{3}$ $\quad$ $\frac{1}{1}$

$\mathfrak{F}_4$: $\frac{0}{1}$ $\quad$ $\frac{1}{4}$ $\quad$ $\frac{1}{3}$ $\quad$ $\frac{1}{2}$ $\quad$ $\frac{2}{3}$ $\quad$ $\frac{3}{4}$ $\quad$ $\frac{1}{1}$

$\mathfrak{F}_5$: $\frac{0}{1}$ $\quad$ $\frac{1}{5}$ $\quad$ $\frac{1}{4}$ $\quad$ $\frac{1}{3}$ $\quad$ $\frac{2}{5}$ $\quad$ $\frac{1}{2}$ $\quad$ $\frac{3}{5}$ $\quad$ $\frac{2}{3}$ $\quad$ $\frac{3}{4}$ $\quad$ $\frac{4}{5}$ $\quad$ $\frac{1}{1}$

$\mathfrak{F}_6$: $\frac{0}{1}$ $\quad$ $\frac{1}{6}$ $\quad$ $\frac{1}{5}$ $\quad$ $\frac{1}{4}$ $\quad$ $\frac{1}{3}$ $\quad$ $\frac{2}{5}$ $\quad$ $\frac{1}{2}$ $\quad$ $\frac{3}{5}$ $\quad$ $\frac{2}{3}$ $\quad$ $\frac{3}{4}$ $\quad$ $\frac{4}{5}$ $\quad$ $\frac{5}{6}$ $\quad$ $\frac{1}{1}$

$\mathfrak{F}_7$: $\frac{0}{1}$ $\quad$ $\frac{1}{7}$ $\quad$ $\frac{1}{6}$ $\quad$ $\frac{1}{5}$ $\quad$ $\frac{1}{4}$ $\quad$ $\frac{2}{7}$ $\quad$ $\frac{1}{3}$ $\quad$ $\frac{2}{5}$ $\quad$ $\frac{3}{7}$ $\quad$ $\frac{1}{2}$ $\quad$ $\frac{4}{7}$ $\quad$ $\frac{3}{5}$ $\quad$ $\frac{2}{3}$ $\quad$ $\frac{5}{7}$ $\quad$ $\frac{3}{4}$ $\quad$ $\frac{4}{5}$ $\quad$ $\frac{5}{6}$ $\quad$ $\frac{6}{7}$ $\quad$ $\frac{1}{1}$

$\mathfrak{F}_8$: $\frac{0}{1}$ $\quad$ $\frac{1}{8}$ $\quad$ $\frac{1}{7}$ $\quad$ $\frac{1}{6}$ $\quad$ $\frac{1}{5}$ $\quad$ $\frac{1}{4}$ $\quad$ $\frac{2}{7}$ $\quad$ $\frac{1}{3}$ $\quad$ $\frac{3}{8}$ $\quad$ $\frac{2}{5}$ $\quad$ $\frac{3}{7}$ $\quad$ $\frac{1}{2}$ $\quad$ $\frac{4}{7}$ $\quad$ $\frac{3}{5}$ $\quad$ $\frac{5}{8}$ $\quad$ $\frac{2}{3}$ $\quad$ $\frac{5}{7}$ $\quad$ $\frac{3}{4}$ $\quad$ $\frac{4}{5}$ $\quad$ $\frac{5}{6}$ $\quad$ $\frac{6}{7}$ $\quad$ $\frac{7}{8}$ $\quad$ $\frac{1}{1}$

Figure 5.25

When the series is extended step by step, the new fractions are mediants of adjacent fractions, which, as we noted earlier, is obtained by adding the numerators and the denominators. That is, if the three fractions $\frac{a}{b}, \frac{p}{q}, \frac{c}{d}$ are three consecutive fractions of a Farey sequence, then the middle fraction $\frac{p}{q} = \frac{a+c}{b+d}$, which is to say that it is the mediant of the two surrounding fractions—that is, $\frac{a}{b} \oplus \frac{c}{d} = \frac{a+c}{b+d}$.

We can also demonstrate the Farey sequence geometrically by taking two circles of radius $\frac{1}{2}$ with centers on a coordinate plane of $\left(0, \frac{1}{2}\right)$ and $\left(1, \frac{1}{2}\right)$, which, in effect, would be tangent to the x-axis and tangent to each other as well. If we then draw a third circle that is tangent to each of these two earlier circles and tangent to the x-axis, we will have constructed a circle whose point of tangency with the x-axis forms a Farey sequence with the first two points of tangency. If we continue this process, we will find that with each additionally "sandwiched-in" tangent circle, the point of tangency with the x-axis adds another member of a Farey sequence.

We can demonstrate this in the following way: we can begin with two circles that are tangent to the x-axis at points $\left(\frac{a}{b},0\right)$ and $\left(\frac{c}{d},0\right)$ and have radii of length $\frac{1}{2b^2}$ and $\frac{1}{2d^2}$, respectively. They will be tangent if and only if $\frac{a}{b}$ and $\frac{c}{d}$ are Farey neighbors, that is if $ad - bc = \pm1$. Furthermore, these two circles will touch the x-axis at the mediant point $\frac{a+c}{b+d}$ and will have a radius of $\frac{1}{2(b+d)^2}$ while being tangent to both of the previous two circles. (See figure 5.26.)

$$r_{\text{left}} = \frac{1}{2b^2}$$

$$r_{\text{right}} = \frac{1}{2d^2}$$

$$r_{\text{middle}} = \frac{1}{2(b+d)^2}$$

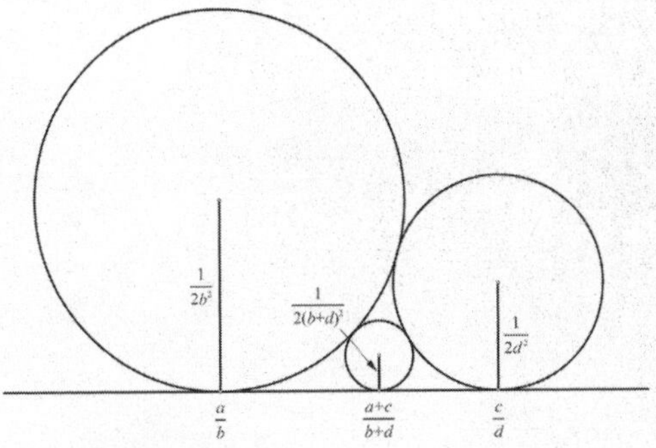

Figure 5.26

In figure 5.26, the two outer circles have radii $r_{\text{left}} = \frac{1}{2b^2}$ and $r_{\text{right}} = \frac{1}{2d^2}$, while the middle circle has radius $r_{\text{middle}} = \frac{1}{2(b+d)^2}$. From this, there is a rather-surprising relationship among the radii of the circles,[8] $\frac{1}{\sqrt{r_{\text{middle}}}} = \frac{1}{\sqrt{r_{\text{left}}}} + \frac{1}{\sqrt{r_{\text{right}}}}$.

This can be justified as follows:

$$\frac{1}{\sqrt{r_{\text{left}}}} + \frac{1}{\sqrt{r_{\text{right}}}} = \frac{1}{\sqrt{\frac{1}{2b^2}}} + \frac{1}{\sqrt{\frac{1}{2d^2}}} = \sqrt{2} \cdot (b+d), \text{ and } \frac{1}{\sqrt{r_{\text{middle}}}} = \frac{1}{\sqrt{\frac{1}{2(b+d)^2}}} = \sqrt{2} \cdot (b+d).$$

A simple justification can also be found in M. Hajja.[9]

Thus, the series of *Ford circles*[10]—named after the American mathematician Lester R. Ford (1886–1967)—provides us with the Farey sequence shown in figure 5.27. This is for the Ford circles for the Farey sequence with $n = 7$:

$$\mathfrak{F}_7: \frac{0}{1}, \frac{1}{7}, \frac{1}{6}, \frac{1}{5}, \frac{1}{4}, \frac{2}{7}, \frac{1}{3}, \frac{2}{5}, \frac{3}{7}, \frac{1}{2}, \frac{4}{7}, \frac{3}{5}, \frac{2}{3}, \frac{5}{7}, \frac{3}{4}, \frac{4}{5}, \frac{5}{6}, \frac{6}{7}, \frac{1}{1}.$$

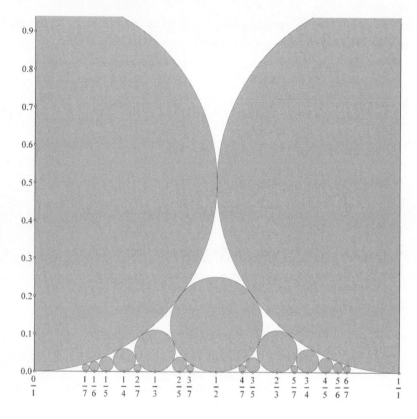

Figure 5.27

Remember, you can test neighboring fractions in the Farey sequence by taking the cross-product and showing that it is equal to ±1. With this unusual arrangement of fractions, we demonstrated that simple fractions yield some rather-unexpected relationships that exist arithmetically, algebraically, and, as we have just shown, geometrically. From this chapter you can see there is far more to fractions and their interrelationships than "meets the eye."

CONCLUSION

We hope that you have enjoyed our journey through various curiosities in mathematics. We have made every effort to keep it simple, and thereby make it more attractive to the general audience. As you will recall, when we began with number relationships in arithmetic peculiarities, we had hoped to set the stage for demonstrating mathematical oddities that unfortunately are often neglected in our learning of mathematics during our school years. It is our belief that by demonstrating some of these unusual aspects of mathematics, more people will be enchanted with this beautiful subject.

There are many ways in which geometry that can be made attractive to the general reader. We chose some that are a bit "off the beaten path." This visual aspect of mathematics has practically endless curiosities that can enchant the intrigued learner. Some of these can be found in our previously published books *Magnificent Mistakes in Mathematics* and *Mathematical Amazement and Surprises*, which show a variety of other peculiarities of this very visual aspect of mathematics.

Problem solving is a key component of mathematics. Naturally, there are countless very challenging mathematical problems that did not guide our selection for inclusion. In chapter 3, we selected ninety problems, each of which has a curious aspect to it. Some were selected for the nature of the question asked—which may seem overwhelming at times. Others were selected to show how a curious solution—atypical in its very nature—can serve as a delightful alternative to the standard and expected method of solution. Also included in this collection are a few tricky problems chosen merely for entertainment. Not to spoil the fun of attempting to solve each of the problems, we deliberately placed the solutions as the second part of that chapter. This way, the temptation to let your eyes wander to the solution has been reduced, since the reader would have to turn pages to get there.

In today's statistically dominated world there is talk about all kinds of central tendency measures. We presented a wide variety of these means and compared their magnitude in various ways. The algebraic comparisons are all very simple using elementary methods, and the geometric comparisons allowed you to "see" their relative magnitudes.

We closed our journey through mathematical curiosities by considering one of the most basic elements in mathematics: fractions. However, we assumed that the reader was proficient in the typical fraction arithmetic, so we showed many unusual applications and aspects of this basic feature of mathematics.

As we hope that you have increased your love for the beauty and power of mathematics, we would expect you to become ambassadors for this most important subject by spreading the word that mathematics not only is important but also can be appreciated for its unusual aspects that have been lost on most students during their school years. Yes, mathematics can be entertaining as well!

NOTES

CHAPTER 1: ARITHMETIC CURIOSITIES

1. Or Nikolaj Petrovič Bogdanov-Bel'skij.

2. See Alfred S. Posamentier and Ingmar Lehmann, *The (Fabulous) Fibonacci Numbers* (Amherst, NY: Prometheus Books, 2007), pp. 27, 97ff.

3. For the interested reader, here is a brief discussion about why this rule works as it does. Consider the number $\overline{abcde}$, where $a, \ldots, e \in \{0, 1, 2, 3, \ldots, 9\}$ and $a \neq 0$, whose value can be expressed as

$$N = 10^4 a + 10^3 b + 10^2 c + 10d + e = (11-1)^4 a + (11-1)^3 b + (11-1)^2 c + (11-1)d + e$$
$$= [11M + (-1)^4]a + [11M + (-1)^3]b + [11M + (-1)^2]c + [11 + (-1)]d + e$$
$$= 11M[a+b+c+d] + a - b + c - d + e, \text{ which implies that divisibility by 11 of } N \text{ depends on}$$

the divisibility of: $a - b + c - d + e = (a+c+e) - (b+d)$, the difference of the sums of the alternate digits. Note: $11M$ denotes a multiple of 11.

4. Repunits are defined as $rn = \dfrac{10^n - 1}{9}$, where $n \in \mathbb{N}^*$. In these numbers, the units-digit 1 repeats n times and are therefore called *repunits* (repeated units). The asterisk indicates the exclusion of zero.

5. Richard L. Francis, "Mathematical Haystacks: Another Look at Repunit Numbers," *College Mathematics Journal* 19, no. 3 (May 1988): 240–46.

6. Wolfram MathWorld, "Repunit," http://mathworld.wolfram.com/Repunit .html (accessed March 23, 2014).

7. This was discovered by Leonhard Euler.

8. Triangular numbers are 1, 3, 6, 10, 15, 21, . . . , and each can represent a number of points that can be arranged to form an equilateral triangle.

9. A *Mersenne number* is a number of the form $Mn = 2n - 1$, where n is an integer. A *Mersenne prime* is a prime number of the form $Mn = 2n - 1$.

10. "Hydroxydeoxycorticosterones" and "hydroxydesoxycorticosterone" have twenty-seven letters.

11. For more on this property, see Posamentier and Lehmann, *(Fabulous) Fibonacci Numbers*.

12. The Lucas numbers, named after the French mathematician Edouard Lucas (1842–1891), are 1, 3, 4, 7, 11, 18, 29, 47, 76, and so on.

13. H. Bonse, "*Über eine bekannte Eigenschaft der Zahl 30 und ihre Verallgemeinerung,*" *Archiv Mathematik-Physik* 3, no. 12 (1907): 292–95.

14. Hans Rademacher and Otto Toeplitz, *The Enjoyment of Mathematics: Selections from Mathematics for the Amateur* (Princeton, NJ: Princeton University Press, 1966), pp. 187–92.

15. Alfred S. Posamentier and Ingmar Lehmann, *Pi: A Biography of the World's Most Mysterious Number* (Amherst, NY: Prometheus Books, 2004), chap. 5, "Pi Curiosities," pp. 137–56.

16. Martin Gardner, *New Mathematical Diversions* (Washington, DC: MAA, 1995).

17. Alfred S. Posamentier and Ingmar Lehmann, *The Glorious Golden Ratio* (Amherst, NY: Prometheus Books, 2012).

18. Those of us who have lived through 1991 and 2002 will be the last generation who will have lived through two palindromic years for over the next one thousand years (assuming the current level of longevity).

19. The definition of $n!$ (spoken as "n factorial") is the product

$n \cdot (n-1) \cdot (n-2) \cdot \ldots \cdot 3 \cdot 2 \cdot 1$. Therefore, $4! = 4 \cdot 3 \cdot 2 \cdot 1 = 24$.

20. $M_p = 2^p - 1$ and p is prime.

21. *Wikipedia*, s.v. "list of perfect numbers," http://en.wikipedia.org/wiki/List_of_perfect_numbers (accessed March 23, 2014).

22. The Mersenne prime exponent n generates them with the expression $2^{n-1}(2^n - 1)$, where $2^n - 1$ is a Mersenne prime.

23. For a proof that this relationship holds as started, see Ross Honsberger, *Ingenuity in Mathematics* (New York: Random House, 1970), pp. 147–56.

24. The nth triangle number is the number of dots composing a triangle with n dots on a side and is equal to the sum of the n natural numbers from 1 to n. The sequence of triangular numbers, starting at the 0th triangular number, is: 0, 1, 3, 6, 10, 15, 21, 28, 36, 45, 55, 66, 78, 91, 105, 120, ... $Tn = 1 + 2 + 3 + \ldots + (n-1) + n = \frac{n(n+1)}{2}$.

25. In case you are interested in seeing how consecutive squares relate to triangular numbers, see Alfred S. Posamentier and Ingmar Lehmann, *Mathematical Amazements and Surprises: Fascinating Figures and Noteworthy Numbers* (Amherst, NY: Prometheus Books, 2009), p. 42.

26. You will find a proof of this statement in B. A. Kordemski, *Köpfchen, Köpfchen!* (Leipzig, Jena, Berlin: Urania, 1963), pp. 202, 318.

27. Alfred Moessner, "Eine Bemerkung über die Potenzen der natürlichen Zahlen," *Sitzungs Berichte Mathematik-Naturwissenschaft Klasse Bayerische Akademie der Wissenschaft*, no. 3 (1952): 29; R. K. Guy, "The Strong Law of Small Numbers," *American Mathematical Monthly* 95, no. 8 (Oct. 1988): 697–712 (examples 17–20, pp. 701–702).

28. Triangular numbers are those that when represented as dots can be arranged to form an equilateral triangle. Oskar Perron, "Beweis des Moessnerschen Satzes," *Sitzungs Berichte Mathematik-Naturwissenschaft Klasse Bayerische Akademie der Wissenschaft* (1951): 31–34; see also Guy, "Strong Law of Small Numbers," pp. 697–712.

29. The factorial of a number is the product of all the natural numbers up to and including that number. For example, $4! = 1 \cdot 2 \cdot 3 \cdot 4 = 24$.

30. Guy, "Strong Law of Small Numbers," pp. 697–712.

31. This was done in 1679 by the German mathematician and philosopher Gottfried Wilhelm Leibniz (1646–1716) when he was fifteen years old.

32. Posamentier and Lehmann, *Mathematical Amazements and Surprises*, pp. 97–104.

33. Verna Gardiner, R. Lazarus, N. Metropolis, and S. Ulam, "On Certain Sequences of Integers Defined by Sieves," *Mathematics Magazine* 29, no. 3 (1956): 117–22.

34. See numbers where all the digits are 1s.

35. Pythagorean triples are three numbers, a, b, and c, in the relationship $a^2 + b^2 = c^2$.

CHAPTER 2: GEOMETRIC CURIOSITIES

1. To be able to solve this task we will assume that Earth is a perfect sphere.

2. Alfred S. Posamentier and Ingmar Lehmann, *Pi: A Biography of the World's Most Mysterious Number* (Amherst, NY: Prometheus Books, 2004), pp. 240–42.

3. Named for the German engineer Franz Reuleaux (1829–1905).

4. Further characteristics of the Reuleaux triangle can be found in Posamentier and Lehmann, *Pi*, pp. 158–70.

5. See Alfred S. Posamentier and Ingmar Lehmann, *The Secrets of Triangles: A Mathematical Journey* (Amherst, NY: Prometheus Books, 2012), pp. 33–42.

348   NOTES

6. *Vie & Oeuvres de Descartes: Étude Historique par Charles Adam*, vol. 4 (Paris: Adam et Tannery/Léopold Cerf, 1901), p. 63.

7. "The Kiss Precise," *Nature* 137, no. 3477 (June 20, 1936): 1,021 and http://www.nature.com/nature/journal/v137/n3477/pdf/ 1371021 a0.pdf (accessed March 31, 2014).

8. Alfred S. Posamentier, *Advanced Euclidean Geometry* (Hoboken, NJ: John Wiley and Sons, 1999), pp. 117–23.

9. Named for Pierre Varignon (1654–1722), a French mathematician.

10. The midline of a triangle is the line segment joining the midpoints of two sides of a triangle.

11. See Posamentier and Lehmann, *Secrets of Triangles*, p. 39.

12. Alfred S. Posamentier and Ingmar Lehmann, *Mathematical Amazements and Surprises: Fascinating Figures and Noteworthy Numbers* (Amherst, NY: Prometheus Books, 2009), p. 176. For a proof of this relationship, see Posamentier, *Advanced Euclidean Geometry*, pp. 121–22.

13. Named after the Russian mathematician Isaak Moiseevich Yaglom [also Isaak Moisejewitsch Jaglom] (1921–1988) and the Italian mathematician Adriano Barlotti (1923–2008).

14. Named after the French mathematician Victor Thébault (1882–1960).

15. Henricus Hubertus van Aubel, later called Henri Hubert (1830–1906), born in Maastricht, Netherlands, died in Antwerp, Belgium.

16. That means a quadrilateral in which the diagonals are perpendicular. In other words, it is a four-sided figure in which the line segments between nonadjacent vertices are orthogonal to each other.

17. Posamentier and Lehmann, *Mathematical Amazements and Surprises*, pp. 177–78.

18. Ibid., pp. 182–83. Posamentier, *Advanced Euclidean Geometry*, pp. 127–32.

19. The Greek title, *Syntaxis Mathematica*, means "mathematical (or astronomical) compilation." The Arabic title, *Almagest*, is a renaming meaning "great collection (or compilation)." The book is a manual of all of the mathematical astronomy that the ancients knew to that time. Book 1 of the thirteen books that represent this monumental work contains the theorem that we presented here and that now bears Ptolemy's name.

20. For proof of Ptolemy's theorem, see Posamentier, *Advanced Euclidean Geometry*, pp. 128–30.

21. For a proof of this relationship, see Posamentier, *Advanced Euclidean Geometry*, pp. 133–34.

22. This problem involves constructing a square with the same area as a given circle—using only straightedge and compass.

23. R. L. Brooks, C. A. B. Smith, A. H. Stone, and W. T. Tutte, "The Dissection of Rectangles into Squares," *Duke Mathematics Journal* 7 (1940): 312–40. In general the partitioning of a polygon P into polygons $P_1, P_2, \ldots, P_n$ is said to be perfect when all the polygons Pj are similar to polygon P and no two are congruent to each other. The number n is said to be the order of the partition.

24. Zbigniew Moroń, "O Rozkladach Prostokatow Na Kwadraty" ("On the Dissection of a Rectangle into Squares"), *Przegląd Matematyczno-Fizyczny* 3 (1925): 152–53.

25. H. Reichardt and H. Toepken, *Jahresbericht der Deutschen Mathematiker Vereinigung* 50 (1940): A 271.

26. C. Müller, "Perfect Squared Squares," (Forschungsergebnisse, N/89/26, Universität Jena, 1989); J. D. Skinner II, *Squared Squares: Who's Who & What's What* (self-published, 1993).

27. See Posamentier and Lehmann, *Secrets of Triangles*, pp. 19–20.

28. For a proof of Desargues's theorem, see Posamentier, *Advanced Euclidean Geometry*, pp. 51–52.

29. Johannes Kepler, *Gesammelte Werke* (Munich, Germany: Beck, 1937), 6:268–69.

30. Ian Stewart, *How to Cut a Cake—and Other Mathematical Conundrums* (New York: Oxford University Press, 2006), pp. 49–51.

31. Alfred S. Posamentier and Ingmar Lehmann, *Magnificent Mistakes in Mathematics* (Amherst, NY: Prometheus Books, 2013).

CHAPTER 3: CURIOUS PROBLEMS WITH CURIOUS SOLUTIONS

1. You will find more examples of this kind in Alfred S. Posamentier and Ingmar Lehmann, *Magnificent Mistakes in Mathematics* (Amherst, NY: Prometheus Books, 2013), pp. 217–21.

2. It takes a person as long to go from his current position to the exit as it would in the opposite direction, from the exit to the current position.

3. H. Steinhaus, *Mathematical Snapshots* (Oxford: Oxford University Press, 1960), p. 59.

4. The quadratic formula for the equation $ax^2 + bx + c = 0$ is $x = \dfrac{-b \pm \sqrt{b^2 - 4ac}}{2a}$.

5. Using the various radii of the congruent circle, we have $\tan\alpha = 2r/6r = 1/3$.

6. The 1878 edition of *The Globe Encyclopaedia of Universal Information* describes loxodrome lines as "a curve which cuts every member of a system of lines of curvature of a given surface at the same angle. A ship sailing towards the same point of the compass describes such a line which cuts all the meridians at the same angle. In Mercator's Projection (q.v.) the Loxodromic lines are evidently straight."

7. Gerhard Mercator (1512–1594)—actual name: Gerard De Kremer; latinized: Gerardus Mercator; German: Gerhard Krämer—was a mathematician, geographer, philosopher, theologian, and cartographer.

CHAPTER 4: MEAN CURIOSITIES

1. For the arithmetic mean, a and b is defined for all real numbers; for the geometric mean, it is defined for $a \geq 0$, and $b \geq 0$; and for the harmonic mean, a and b are defined for all real numbers $a,b > 0$ and $a + b = 0$.

2. A geometric sequence is one with a common factor between consecutive terms.

3. Also referred to as the *quadratic average*.

4. E. Maor, "A Mathematician's Repertoire of Means," *Mathematics Teacher* 70 (1977): 20–25.

5. Heinz Bauer, "Mittelwerte und Funktionalgleichungen," *Sitzungsberichte der Bayerischen Akademie der Wissenschaften. Mathematisch-Naturwissenschaftliche Klasse* (1986), Munich, (1987): 1–9; Helmut Titze, "Zur Veranschaulichung von Mittelwerten," *Praxis der Mathematik* 29, no. 4 (1987): 200–202.

6. We use the fact that if a line parallel to one side of a triangle contains the midpoint of a second side, then it will also contain the midpoint of the third side.

7. Since in the right triangle on the x-axis $xS\ (= xM)$, we have $OS^2 = xS^2 + yS^2$
$= xM^2 + yR^2 = xM^2 + (OR^2 - xR^2) = (a ⓐ b)^2 + [(a ⓐ b)^2 - (a ⓖ b)^2] = 2(a ⓐ b)^2 - (a ⓖ b)^2$
$= 2 \cdot \left(\dfrac{a+b}{2}\right)^2 - \left(\sqrt{ab}\right)^2 = \dfrac{(a+b)^2}{2} - ab = \dfrac{a^2 + 2ab + b^2}{2} - \dfrac{2ab}{2} = \dfrac{a^2 + b^2}{2}$.

8. Naoki Sato, "Three Gems in Geometry," *Crux Mathematicorum* 25, no. 8 (1999): 498–501; Howard Eves, "Means Appearing in Geometric Figures,"

Mathematics Magazine 76, no. 4 (2003): 292–94; Peter S. Bullen, *Handbook of Means and Their Inequalities* (Dordrecht, Boston, London: Kluwer Academic Publishers, 2003).

CHAPTER 5: AN UNUSUAL WORLD OF FRACTIONS

1. A *harmonic sequence* is a sequence of numbers whose reciprocals form an arithmetic sequence—a sequence with a common difference between terms. When the same-quality strings have lengths that represent a harmonic sequence and are strummed together, the sound emitted is said to be harmonious.

2. *Tetrahedral numbers* are the sequence of partial sums of the triangular numbers:

1, $1+3=$**4**, $1+3+6=$**10**, $1+3+6+10=$**20**, $1+3+6+10+15=$**35**, $1+3+6+10+15+21=$**56**, ..., $\frac{n(n+1)(n+2)}{6}$, ...

Also see Alfred S. Posamentier and Ingmar Lehmann, *Mathematical Amazements and Surprises: Fascinating Figures and Noteworthy Numbers* (Amherst, NY: Prometheus Books, 2009), p. 42.

3. *Pentatope numbers* are the partial sums of the oblique of tetrahedral numbers as shown here:

1, $1+4=$**5**, $1+4+10=$**15**, $1+4+10+20=$**35**, $1+4+10+20+35=$**70**, $1+4+10+20+35+56=$**126**, ..., $\frac{n(n+1)(n+2)(n+3)}{24}$, ...

4. See J. H. Conway and R. K. Guy, *The Book of Numbers* (New York: Springer, 1996).

5. Laurence Sherzer, "McKay's Theorem," *Mathematics Teacher* 66 (1973): 229–30. See also Thomas E. Kriewall, "McKay's Theorem and Farey Fractions," *Mathematics Teacher* 68 (1975): 28–31.

6. J. Farey, "On a Curious Property of Vulgar Fractions," *Philosophical Magazine* 47 (1816): 385–86.

7. Haros was a geometer (mathematician) in the French Bureau du Cadastre at the end of the eighteenth century and the beginning of the nineteenth century. He is best known for a small table he prepared to convert fractions to their decimal equivalents. Applying Chuquet's algorithm, he developed a series of mediants that the English mathematician Henry Goodwyn (1745–1824) used to create a table of decimal equivalents—something likely known to Farey.

8. R. Honsberger, *Mathematical Gems* (Washington, DC: MAA, 1973), pp. 52–53, 153–55; and R. Honsberger, *Mathematical Morsels* (Washington, DC: MAA, 1978), pp. 218–19.

9. M. Hajja, "On a Morsel of Ross Honsberger," *Mathematical Gazette* 93, no. 527 (2009): 309–12.

10. L. R. Ford, "Fractions," *American Mathematical Monthly* 45, no. 9 (1938): 586–601. See also Conway and Guy, *Book of Numbers*, pp. 152–54.

INDEX

Page numbers in *italic* indicate solutions for word problems in chapter 3.

NAMES AND SUBJECTS

smallest natural number that can be
expressed as the sum of cubes of
natural numbers, 23
in a triangle of natural numbers find
the row where 2000 can be found
(Problem 65), 196–97, *263*
and unit fractions, 316
negative roots, 137, 308
show that $x^4 - 5x^3 - 4x^2 - 7x + 4$
$= 0$ has no negative roots
(Problem 30), 181, *230*
neighbor fractions, 336, 337
Newton, Isaac, 317
New York Herald Tribune (newspaper),
61
Nicomachus, 60, 95
North Pole (Problem 45), 186, *243–45*
numbers
cyclic number loop, 73–76
numerical sequences, 25, 102–103,
106, 108, 346n24
$192 + 384 = 576$, 51
fraction sequences, 330, 331,
337–42, 351n1
and means, 292, 294, 300, 317,
322, 324, 328, 350n2
multiplication that uses all nine
non-zero numbers, 65
next number in sequence 1, 2, 3,
4, 6, 8, 9, 12, 16, 18, 24, 27,
32, 36, 48, 54, 64, 72, 81, 96
. . . (Problem 86), 205,
285–86
pentatope numbers, 331, 350n3
Polignac numbers, 94–95
representing numbers 1 to 100 using
only 4s, 86–92
smallest number

divisible by each of the nine
digits in base-10 system
(Problem 9), 174, *213*
to meet criteria (eggs in a
basket) (Problem 24), 180,
227
prime number consisting of ten
different digits (first digit not
a zero) (Problem 26), 180,
227–28
symmetric numbers, 51
See also composite numbers; decom-
position of numbers; even numbers;
factorials; Fibonacci numbers; four-
digit numbers; integers; Kaprekar
numbers; Lucas, Edouard, and
Lucas numbers; manipulating
numbers; natural numbers; odd
numbers; patterns in numbers;
perfect numbers; prime numbers;
repunits; tetrahedral numbers; three-
digit numbers; triangular numbers;
two-digit numbers; units digit; zeros
numerators
canceling same numerator and
denominator to reduce a fraction,
15–17, 18–19
converting fraction to decimal form,
82
mediants, 334–35, 340
unit fractions having 1 as, 12, 317
Numerorun Mysteria (Bungus), 60

octagon, fraction of the area of an
octagon is area of shaded isosceles
triangle inside (Problem 85), 204–
205, *284–85*

ten different digits (first digit not a
zero) (Problem 26), 180, *227–28*
that create new prime numbers with
each arrangement of their digits, 30
probability problems
of a point in a larger circle will be in
a smaller circle inscribed within
(Problem 195), 194–95, *259–60*
of two classmates having same
birthdate (Problem 70), 199,
268–70
problem solving, 11–12, 343
mathematical problems and solutions, 171–289
See also logical-thinking problems
Ptolemy (Claudius Ptolemaeus) and
Ptolemy's theorem, 157–58, 348n19
Pythagoras, 291
Pythagorean means, 291. *See also*
arithmetic mean; geometric mean;
harmonic mean
Pythagorean theorem, 118–20, 151, 301
not extending beyond the power of
2, 121–22
Ptolemy's theorem applied to rectangles establishing, 157–58
role of in the rope-around-the-
equator exercise, 128
Pythagorean triples, 111, 347n35

quadratic average, 350n3. *See also*
root-mean-square
quadratic formula for equation: ax^2
$+ bx + c = 0$, 229, 350n4
quadrilaterals
creating a parallelogram from midpoints on a quadrilateral, 145–48
cyclic quadrilaterals, 155–58

having two centers, 148–49. *See also*
centerpoint of a quadrilateral; centroids of a triangle and a quadrilateral
isosceles right triangle and square,
determining size of the shaded
quadrilateral formed (Problem
10), 175, *214*
lengths of two sets of diagonals,
150–52
orthodiagonal quadrilateral, 154,
348n16
placing squares on the sides of a
parallelogram, 152–55
ratio of area of quadrilateral to the
area of a triangle (Problem 67),
197–98, *264–65*
using to compare means, 303–305
quotient. *See* division, problems
involving

Rademacher, Hans, 37
radiator needing water, measuring
amount for (Problem 58), 193–94,
256–57
radical expressions with higher roots,
which is larger (Problem 71), 200,
270–71
radius, length of line equal to radius of
a semicircle (Problem 39), 184–85,
237–38
Raleigh, Walter, 167
ratios, 192
ratio of area of quadrilateral to the
area of a triangle (Problem 67),
197–98, *264–65*
ratio of areas and sides of two equilateral triangles inscribed and